Managing the Asian Century

Series editor

Purnendu Mandal, College of Business, Lamar University, Beaumont, TX, USA

Managing the Asian Century provides a platform for scholastic discussions and stresses the need for a holistic framework to understand Asia as an emerging economic global powerhouse. Books published in this series cover Asia-centric topics in economics, production, marketing, finance, entrepreneurship, education, culture, technology, as well as other areas of importance to Asian economics. The series will publish edited volumes based on papers submitted to international and regional conferences that focus on specific Asia-Pacific themes, such as investment in education, women's rights, entrepreneurship, climate change, wage inequality, challenges in governance, and corruption. Books in this series are of keen interest to researchers and policy planners around the world and will be used by universities for graduate and doctoral level studies.

More information about this series at http://www.springer.com/series/13579

J. Maiti · Pradip Kumar Ray
Editors

Industrial Safety Management

21st Century Perspectives of Asia

Springer

Editors
J. Maiti
Indian Institute of Technology Kharagpur
Kharagpur
India

Pradip Kumar Ray
Indian Institute of Technology Kharagpur
Kharagpur
India

ISSN 2364-5857 ISSN 2364-5865 (electronic)
Managing the Asian Century
ISBN 978-981-10-6327-5 ISBN 978-981-10-6328-2 (eBook)
https://doi.org/10.1007/978-981-10-6328-2

Library of Congress Control Number: 2017950022

Printed on acid-free paper

This Springer imprint is published by Springer Nature
The registered company is Springer Nature Singapore Pte Ltd.
The registered company address is: 152 Beach Road, #21-01/04 Gateway East, Singapore 189721, Singapore

Preface

The International Conference on Management of Ergonomic Design, Industrial Safety and Health Care Systems (MESH 2016) was held during 20–23 December 2016, at the Department of Industrial & Systems Engineering (ISE), Indian Institute of Technology Kharagpur, West Bengal, India. The focus of the conference was on three interrelated domains: (i) Ergonomic Design, (ii) Industrial Safety, and (iii) Healthcare Systems. Of them, the selected articles under 'Industrial Safety' have been included in this volume. The articles are categorized under three themes: (i) Safety by Design, (ii) Safety Analytics, and (iii) Safety Management.

'Safety by Design' is basically a concept of applying methods to minimize hazards early in the design process, with an emphasis on maximizing employee health and safety throughout the life cycle of products, materials, and processes. This concept can be applied in various areas like improving safety and productivity at the construction site (*See Vigneshkumar and Maheswari, page 1–10*) reducing risk associated with EOT crane's operations through virtual prototypes (see Dhalmahapatra et al., page 11–25), reduction of concentration of coal dust in working environment by using water mist systems (*see Vivek and Manikandan, page 26–36*), and design of antilock braking system as per rough road conditions (*see Vivekanandan and Fulambarkar, page 37–51*).

'Safety Analytics' that deals with data-driven decision making can be used in various areas like analysing safety performance and pointing out the specific areas of improvement by mapping safety factors based on safety data and incident reports (see Verma et al., page 52–62), predicting occupational incidents (*see Sarkar et al., page 63–78*), evaluating the occupational hazards and their contribution to the occurrences of injuries in hard rock mines (*see Sarkar et al., page 79–93*), modelling of human energy consumption of workers who are repeatedly exposed to vibrations during machining operations (*see Mohod and Mahalle, page 94–104*), and road safety (*see Srinath et al., page 105–115*).

'Safety Management' refers to managing business activities and applying principles, framework, processes to help prevent accidents, injuries, and to minimize other risks. 'Safety Management' concepts are used to develop a framework for implementing the safety life cycle management (SLCM) approach for a safety instrumented system (SIS) to manage the plant safety (*see Rohit et al., page 116–130*), and to understand human–computer interaction behaviour for various systems (*see Rahman, page 131–139*).

After reading this volume, readers will be able to understand the concept and issues related to industrial safety, induction of safety at the design stage to improve the safety performance, analysing, predicting, and reducing hazards by the use of analytics and safety management.

As organizers of MESH 2016, we would like to express our sincere thanks to Director, Prof. Partha Pratim Chakrabarti, IIT Kharagpur, for his overall support and encouragement and Dean (CEP), IIT Kharagpur, for the administrative support in conducting the conference. We are extremely thankful to Tata Steel Limited, Viz experts, Tobii Pro Vitasta India, TVS, Linde, and Janatics Pneumatic for sponsoring the conference/ events. We would also like to thank the members of the *National and International Advisory Committee* for their guidance, the members of the *Technical Committee* and reviewers for reviewing the papers, and the members of the *Organizing Committee* for organizing the entire events in conference. We are especially grateful to the proceedings publisher Springer for publishing the proceedings in the prestigious series of "*Industrial Safety Management*". Moreover, we would like to express our heartfelt appreciation to the plenary speakers, session chairs, and student members. In addition, there still remain an ample number of colleagues, associates, friends, and supporters who helped us in immeasurable ways, and without their support, we would not have achieved a grand success in conducting MESH 2016. Finally, we would like to thank all the speakers, authors, and participants for their contributions that made MESH 2016 successful and all the hard work worthwhile.

We also do believe that the articles within this volume will be useful for the researchers pursuing research in the field of industrial safety, occupational health, and related areas. Practicing technologists would also find this volume to be an enriched source of reference.

Kharagpur, India

J. Maiti
Pradip Kumar Ray

Contents

Part I
Safety by Design

Chapter 1
Prevention Through Design: A Concept Note for Preventing Accidents/Injuries to Construction Workers

C. Vigneshkumar and J. Uma Maheswari

Abstract Planning for health and safety in construction has always been a challenge. Workers at construction sites are exposed to a variety of health hazards every day. Researchers had identified that the different sources of health hazards at construction site include chemical, physical, biological, and ergonomic hazards. Hazardous activities in construction project such as demolition works, excavation works, scaffolding and ladder works, construction machinery, and tools usage cause harm to workers at site which may result in musculoskeletal disorder, respiratory diseases, and dermatitis to workers. Traditionally, construction safety is not considered by designers/architects in project design. Therefore, planning for safety and health of the workers is very critical. The purpose of this paper is to introduce the concept of "Prevention through Design (PtD)" for construction workers, which integrates health and safety into the management of construction projects. The concept of PtD is applicable to diverse disciplines such as construction, health care and social assurance, forestry and fishing, transportation, mining, agriculture, and manufacturing. In this paper, different sources of health hazards and their effects on construction workers were discussed. Further, this paper describes how hazards in construction projects can be prevented or minimized by integrating PtD. This concept is expected to improve the performance of safety and productivity in construction.

Keywords Construction projects · Prevention through design · Health and safety
Hazards

C. Vigneshkumar (✉) · J. Uma Maheswari
Department of Civil Engineering, Indian Institute of Technology Delhi, New Delhi, India
e-mail: civil.vigneshkumar@gmail.com

J. Maiti and P.K. Ray (eds.), *Industrial Safety Management*,
Managing the Asian Century, https://doi.org/10.1007/978-981-10-6328-2_1

1.1 Introduction

1.1.1 Study Background

Construction industry is one of the most hazardous industries all over the world. The overall days lost by injuries and illness of construction were higher than any other industry sector. About 3% of construction workers in Great Britain suffered from work-related illness which leads to 1.2 million days lost (OSH 2004). Annually around 69,000 workers were suffering from illness in Great Britain. Mitigating the hazards is one of the challenging issues in construction projects. Identifying the root causes of hazards plays major role in many construction activities. Different companies from different industries put effort individually and jointly to improve safety in construction, but their results remain unchanged. Particularly, the construction industries in developing countries are found to be most dangerous on safety criteria. For past decades, the academics and professionals from different specialization worked jointly to address safety issues in construction. However, the results which they obtained were not sufficient to prevent hazards in construction. At present, the researchers have focused on the concept of preventing or minimizing the hazards in construction through design. Many researchers stated that addressing workers safety in the design phase of project will improve safety performance in construction. But in many countries, workers' safety is not a part in designer's role.

Design has the major impacts on construction safety (BLS; Driscoll et al. 2008; Kamardeen 2013; Behm 2005; Smallwood 2008) Australia National Coroner's Information System has drawn the conclusion that 37% of 210 identified workplace fatalities had design related issues (Driscoll et al. 2008). Nearly, 50% of construction workers in South Africa thought construction safety is impacted by design (Smallwood 2008). The analysis of data obtained from National Institute of Occupational Safety and Health (NIOSH) Fatality Assessment Control and Evaluation (FACE) program show that 42% of the 224 fatalities were related to design (Behm 2005). About 16.5% of construction accidents were design related incidents. The collected data from seven different countries were analyzed, and the result shows that 35% of construction accidents were related to design (Kamardeen 2013).

1.1.2 Health Hazards in Construction

Construction workers are exposed to a variety of health hazards every day. A hazard is a situation that can cause harm to people, environment, life, and property. In construction, health hazard is something that can cause harm to workers health. Hazardous activities which can cause harm to workers in construction include demolition works, excavation works, scaffolding and ladder works, construction machinery, and tool usage. The hazardous substances associated with construction works are asbestos, lead, silica dust, gases, and fungi. Effect of this substances may result in acute injury, chronic illness and permanent disability or even death to workers. Exposure may differ from day to day, time to time, or even task to task. Chronic health effects develop slowly, whereas acute health effects can be seen quickly. For example, if the worker is affected by fungi during housekeeping, it may cause immediate itchiness and skin irritation. Extreme noise may lead to hear loss temporarily or permanently. In chronic health effect, if the worker breathes a small amount of silica dust, it will not affect the worker immediately but if the worker inhale silica dust regularly, then the worker is exposed to silicosis. The different source of health hazards at construction site includes chemical, physical, biological, and ergonomic hazards.

Chemical hazards are often airborne which can appear as gas, vapor, fume, dust, and mist. For example, pneumatic breakers, tunnel operators, drillers, and masons during breaking and crushing of stone/concrete/bricks get exposed to silica dust and suffer from silicosis. Welders and flame cutting operators during cutting and dismantling tanks get bronchitis. Building demolition works and steam pipe fitters get exposed to asbestos and suffer from asbestosis. Painters and others those who come in contact with solvents get neurological disorder. Workers who use materials which contain epoxy resins, acrylic resins, nickel, cobalt, timber will suffer from allergic dermatitis.

Physical hazards are due to exposes like extreme temperatures, noise, vibration, and radiation which cause different health hazards. Activities such as demolition, drilling, welding cause noise-induced hearing loss. Pneumatic breakers, disk grinders, hand tools cause vibration-induced carpal tunnel syndrome which affects fingers and hands of workers. Exposure to extreme heat and cold causes heat rashes, heat stroke, white finger, etc. Ionizing radiation affects the workers who come in contact with radioactive substance which are previously stored or used in site or may be released during demolition. Those who work under radioactive substance suffer from genetic disease and even cancer.

Biological hazards are due to infectious microorganisms or animal/insect attack at site. Workers are exposed to a risk of bacteria, poison plant, fungi, etc., while performing activities like housekeeping, excavation, site clearing. This may cause immediate itchiness and skin irritation to workers who perform the tasks.

Ergonomic hazards are due to lifting and carrying of objects, repetitive works, working in awkward postures, forceful and muscle efforts, external pressure. Workers who perform concrete work, flooring, roofing, painting, welding, and housekeeping are exposed to high risks of ergonomic threats. Workers who come in contact with tools and sharp objects feel external pressure. Activities such as drilling, hammering, painting with brush pose repetitive works. Workers who perform flooring, drywall insulation, masonry works, welding are used to work in awkward postures. The overall ergonomic threats cause musculoskeletal disorder leading to lifetime pain and/or body disability.

1.1.3 Research Study

Traditionally construction safety is not considered by designers/architects in project design. The construction safety will be improved, if the construction architects/ design engineers are responsive to safety consequences for their design decisions (Gambatese et al. 1997). Since designers do not have required knowledge and experience for playing a role on safety, they are not aware of their impact on safety. Identifying and eliminating risks at the design stage can achieve improved productivity and reduction in environmental damage (Hinze and Wiegand 1992). Certain measures must be performed at the beginning phase of design to reduce the possibility of incidents during construction (Yang and Li 2015). According to the researcher's conclusion, it can be deduced that hazards and risks associated with construction projects can be prevented through design if designers are in a position for decision making.

In this paper, an attempt has been taken to reduce or minimize the health hazards in construction through design. The concept of PtD was introduced which helps to address the workers' safety and health in design phase of the construction projects. Different sources of health hazards and their effects on construction workers are discussed. Further, this paper describes how hazards in construction projects can be prevented or minimized by integrating PtD. Conclusion drawn in this paper will help the construction actors to improve the performance of safety and productivity in construction projects.

1.2 Concept of Prevention Through Design

One of the best way of preventing occupational injuries, illness, and fatalities is to "design-out" the hazards and risks, thereby eliminating or minimizing them during design process. In 1955, the idea of prevention during the design phase was first addressed in the National Safety Council's prevention manual. US industry did not begin the application of PtD concepts until Construction Industry Institute sponsored (Gambatese et al. 1997). Later, the concept of PtD was slowly recognized and applied in the USA (Yang and Li 2015). To promote this concept, National Institute for Occupational Health and Safety (NIOSH) called all major industries in 2007 and launched a national initiative called "Prevention through Design" to highlight hazard mitigation during design stage of tools, equipment, and work processes. Many companies in the USA start supporting the concept of PtD and actively promote them as well. Architects and project owners of many companies in the UK starts addressing the safety in design phase of project, and as a result, many companies responded the changes in management practice. Followed by, Australia set "eliminating hazards at the design stage" as one of the five national priorities in Australian National Occupational Health and Safety Strategy 2002–2012.

For the development of a national initiative on PtD, NIOSH partnered with organizations that include American Industrial Hygiene Association (AIHA), American Society of Safety Engineers (ASSE), Center to Protect Worker's Rights (CPWR), Kaiser Permanente, Liberty Mutual, National Safety Council (NSC), Occupational Health and Safety Administration (OSHA), ORC Worldwide and Regenstrief Center for Healthcare Engineering. To eliminate or reduce the hazards associated with work, PtD concept involves the design of tools, equipment, and work processes. The PtD workshop was first held at Washington in July 2007, which aims at eliminating occupational hazards and controlling risks at the early stage of project life cycle. Approximately 225 participants were attracted from diverse industrial sectors and the technical papers authored by experts in PtD; break-out session reports from industries were published as a special edition of Journal of Safety Research in 2008 (Yang and Li 2015).

The concept of PtD is applicable to diverse disciplines of agriculture, mining, transportation, forestry and fishing, construction, health care and social assistance, warehousing, and manufacturing for preventing workers from hazards through design. The concept of Prevention through design is defined as "*Addressing occupational safety and health needs in the design process to prevent or minimize the work-related hazards and risks associated with the construction, manufacture, use, maintenance, and disposal of facilities, materials, and equipments*" (Ertas 2010). The goal of PtD is to reduce occupational injuries and illness by integrating PtD process in all stages of generic design process. Figure 1.1 shows the process of PtD, which starts with defining the work related to the product and then identifying and analyzing the hazards associated with the products and finally establishing the control measures if hazards cannot be eliminated.

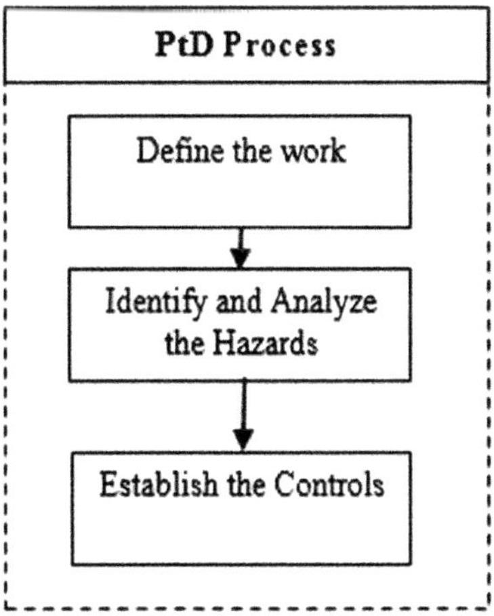

Fig. 1.1 Process of PtD (*Source*: *Prevention through design: Transdisciplinary process*)

1.3 PtD for Construction

PtD is a process of integrating the concept of health and safety into the management of construction projects. PtD for construction helps in addressing safety and health of workers in design process to prevent or minimize the hazards associated with projects. The main goal of PtD is to reduce occupational injuries and illness by integrating PtD process in all stages of generic design process of project (Deborah and Young-Corbett 2014). Figure 1.2 shows the integration of PtD process in generic design process of construction project.

Generic design process starts with defining clients requirements. In this stage, the client's requirements for the projects are described by the design team. The design team will closely work with clients to determine the functional and design requirements for the projects. This stage will serve as a foundation for next upcoming stages. In concept design stage, viable project solutions are identified, through which the best optimum approach can be selected. Followed by, feasibility of selected concept is accomplished to ensure that the concept is achievable both technically and within cost constraints. The main objective of the concept design stage is choosing the alternative feasible concept through which the health and safety of workers can be protected from hazards. The research team members who involved in feasibility stage of design process must have experience and good judgment in decision making. Preliminary design stage bridges the gap between

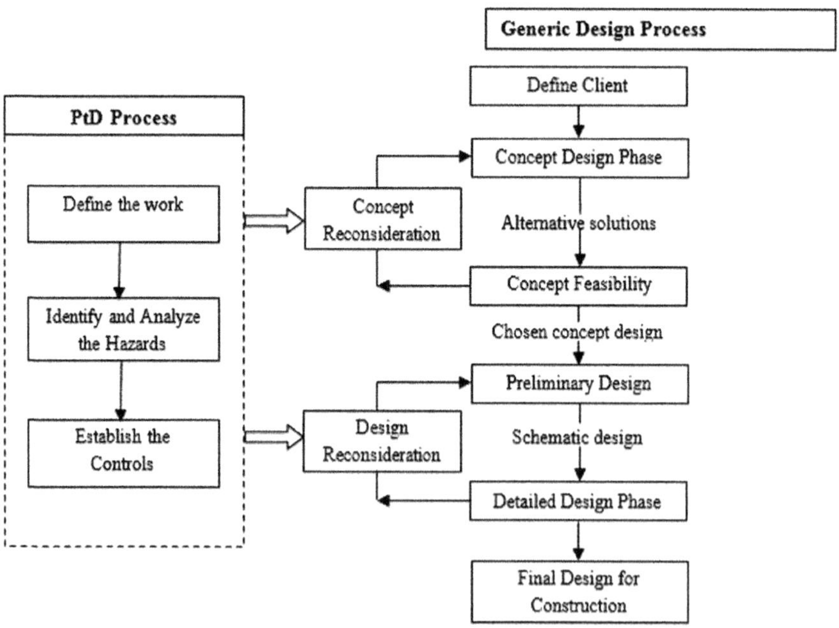

Fig. 1.2 Integration of PtD process in generic design process

concept design stage and detailed design stage. During preliminary design stage, the concept developed is further defined. If more than one concept is involved in concept design stage, the assessment over best solution is selected in preliminary design stage. In preliminary design stage, the concept development must be incremental rather than complete re-examination. The overall system configuration includes schematic and drawings, diagram, layouts, and other engineering configurations which are defined during preliminary design stage. Further, the safety consideration and selection of control measures must also be checked in preliminary design stage to know whether it is still appropriate for hazard analysis. Once the preliminary design is set, the detailed design phase of the project will be prepared with a detailed specification of each component. In the detailed design stage, a final set of hazard analysis is developed which will serve as a basis for the final design for construction.

1.4 PtD for Health Hazards in Construction

The main exposing activities are those which are associated with health hazards of construction projects such as excavation, manual handling, housekeeping, demolition, and concrete/brick work. The workers who perform these tasks are repetitively exposed to different kinds of health hazards. PtD process "design-out" hazards at workplace by integrating PtD in generic design process of projects. Some engineering control options are available to reduce expose to hazards at site. The process of integration of PtD in design process will select/allow to use the materials which are less hazardous, for example, solvent-free or low-solvent adhering and water-based paints. The advantage of integrating PtD in design process is that the designers can come to know about the issues in activities and alert the workers who perform the tasks.

The process which creates vibration can be minimized by selecting alternative process. For example, vibrators which are used manually can be replaced by machine vibrators. To keep workers away from microorganisms, workers must be allowed to use tools/equipment rather than working with hand during site clearing. The well-trained workers are allowed to do housekeeping to avoid unnecessary movements. The process which creates noise and dust can be avoided through choosing alternative process. For example, the methods which produce noise and dust during concrete/bricks breaking can be minimized by choosing alternative method.

To avoid musculoskeletal disorder, alternative equipment or methods must be used where workers are not able to carry or lift manually. The tools which are used by the workers of specific tasks must be less in weight and easy to access.

The demolition and renovation work contains more health hazards, and it can be avoided by choosing alternative process. For example, the noise produced during demolition work can be minimized by using alternative equipment or tool to demolish. The administrative controls and PPEs are used when hazards in sites are not able to prevent by engineering controls.

1.5 Conclusions

Health and safety issues are increasing day by day in construction site. Construction workers are exposed to chemical, physical, biological, and ergonomic hazards that may result in acute injury, chronic illness and permanent disability or even death to workers. Identifying and mitigating the hazards in construction is one of the challenging roles for many construction actors. Traditionally, construction safety is not considered by designers/architects in project design.

PtD is a process of integrating the concept of health and safety into the management of construction projects. In this paper, the concept of PtD is used to prevent workers from hazards in construction worksite. There is a promise from the researches that PtD eliminates or reduces the hazard in the design phase of the hierarchy of control solutions. Therefore, addressing PtD in construction industry will improve the performance of construction safety and decrease the injuries rate. The concept of PtD suggests that identifying and eliminating risks at design stage will reduce accident rate and improve productivity in construction. Further research is needed to find how alternative methods for different activities will improve safety in site.

References

Behm, M.: Linking construction fatalities to the design for construction safety concept. Saf. Sci. **43**, 589–611 (2005)

BLS (Bureau of Labour Statistics). http://www.bls.gov/iif/oshcfoil.htm

Deborah, E.Y.-C.: Prevention through Design: health hazards in asphalt roofing. J. Constr. Eng. Manag. **140**(9), 4 (2014)

Driscoll, T.R., Harrison, J.E., Bradely, C., Newson, R.S.: The role of design issues in work-related fatal injury in Australia. J. Saf. Res. **39**, 209–214 (2008)

Ertas, A.: Prevention through design: transdisciplinary process, The Academy of Transdisciplinary Learning and Advanced Studies. Texas, 1–52 (2010)

Gambatese, J.A., Hinze, J.W., Haas, C.T.: Tool to design for construction worker safety. J. Archit. Eng. **3**, 32–41 (1997)

Hinze, J., Wiegand, F.: Role of designers in construction worker safety. J. Constr. Eng. Manag. **118**(4), 677–684 (1992)

Kamardeen, I.: Design for safety analysis support system for building designers. In: Proceedings of the 13th International conference on construction applications of virtual reality, pp. 56–63. London Press, UK (2013)

OSH (Occupational Safety and Health), "Guidance notes on Health Hazards in Construction Work" (2004) http://www.labour.gov.hk/eng/public/oh/OHB82.pdf

Smallwood, J.: The influence of architectural designers on construction ergonomics. J. Ergon. Soc. S. Afr. **20**, 40–55 (2008)

Yang, L., Li, Q.M.: The research review of the impact of design to construction safety. Front. Eng. Manag. pp. 224–227 (2015). doi:10.15302/J-FEM-2015044

Chapter 2
Virtual Prototype based Simulator for EOT Crane

Krantiraditya Dhalmahapatra, Souvik Das, Sagar Kalbande and J. Maiti

Abstract Electric overhead travelling (EOT) crane is associated with complex operational activities such as loading and unloading in manufacturing industry. Training to operators in real scenario is not possible because of these complex activities and the risk associated with the operation. It may also hamper the manufacturing process of the organization. This study aimed at developing a virtual prototype-based EOT crane simulator that can be used for providing training to the operators. In the proposed approach, EOT crane is considered as 3-degree-of-freedom system, and all the components related to the EOT crane operational environment have been modelled using 3D modelling software. All the models are then integrated in a single layout, and the virtual environment of the EOT crane has been made using unreal engine software. The operational process of the EOT crane can be simulated using keyboard and mouse of the computer in this virtual environment. Organization can adopt this simulator for giving training to their operators and to make the training process more time efficient and cost-effective.

Keywords Virtual reality · 3D modelling · Game engine

K. Dhalmahapatra (✉) · S. Das · S. Kalbande · J. Maiti
Department of Industrial & Systems Engineering, IIT Kharagpur, Kharagpur, India
e-mail: kranti.dhalmahapatra@gmail.com

S. Das
e-mail: rndas9@gmail.com

S. Kalbande
e-mail: sagarkalbande1993@gmail.com

J. Maiti
e-mail: jhareswar.maiti@gmail.com

J. Maiti and P.K. Ray (eds.), *Industrial Safety Management*,
Managing the Asian Century, https://doi.org/10.1007/978-981-10-6328-2_2

2.1 Introduction

Virtual reality is an organized and simulated environment. It is an interactive environment where any user can observe his action and responses in the so-called virtual environment (VE) through some human–computer interface equipment. The mode of interaction between user and VE using some input devices like keyboard, mouse or some multimodal devices like multidirectional treadmill, wired glove, polhemus boom arm, joystick, trackball, exoskeleton (external hardware on hands), eye tracking, video analysis, brain wave electroencephalography (EEG), electromyography (EMG) help in perception of the real-time virtual 3D objects and subsequent analysis of the object.

Virtual reality is highly flexible because construction and visualization of different models are possible according to requirements. VR allows immersion, i.e. user can storm inside the data, interact with objects and manipulate, reshape dimension and scales according to preferences. VR is a dynamic experience because users play the role of both user and observer. Virtual reality consists of three levels: first one is non-immersive, i.e. virtual creation of 3D objects in the computer system without use of any special purpose hardware system. This is useful for training purposes. Flight simulator is a suitable example for this level of VE. Second one is semi-immersive or sensory immersive which has wide applications in robot navigation, airplane simulation and construction modelling where an user can find himself in the virtual environment and perform the necessary action according to his own perception to simulate reality. The last but not the least is fully immersive or neural-direct where connection lies between human brain, database and the user's location and orientation. In this level, a sensory input was imparted to the human brain and simultaneously user's thought process is injected to the virtual environment for visualization and manipulation.

In recent years, production safety scenario is not optimistic; hoisting and conveying machinery operation accidents occur from time to time. The primary reasons for these accidents include the operators' lack of professional knowledge, lack of skills, inadequate standard operating procedure. At present, the crane is recognized as special equipment for heavy operations, but the training process is lagging behind OSHA Safety Training Courses (retrieved from http://www.bebsoft.com/osha-safety-training). Inadequate training material leads to unorganized training of crane operators and they do not really acquire the practical feel of the real-time operation, leaving a potential for accident risk. Therefore, it is an urgent task to change the current training method of the crane operators, improve the professional quality and operation skills of the workers, and ensure the safety and stability of crane operations. At present, virtual reality technology has been applied in many fields such as medical, military, aviation, robotics and manufacturing, construction, education and entertainment. It will also have a significant effect on improving the training of crane operators.

2.2 Virtual Prototype

Virtual prototyping (VP) is creation of simulated environment of a product where analysis and testing of the product can be done for design or qualitative performance analysis. It will help in product development owing to its availability and low prices (Choi et al. 2015) along with replacing the problems such as deployment of dedicated equipment and space in case of conventional training approaches (Goulding et al. 2012). In safety domain of a manufacturing industry, VP can be useful in solving the issues related to layout, operation and dynamic analysis of operating environment (Guo et al. 2013).

In order to change the status quo of the crane operator training, development of an overhead crane operation simulation system based on virtual reality technology is our goal. The developed simulated VR environment can be used for overhead crane operation training, skill improvement training, safety education and practice examination. The simulated system of overhead crane is composed of several parts, including visual system, human–computer system and hardware system ("The Virtual Crane" retrieved from http://venus.web.cern.ch/VENUS/vcs/virtcra2.2).

The use of virtual reality for simulating EOT crane operation improves the understanding of the mechanical structure and working principle of crane system. System architect uses 3D virtual simulation technology to produce realistic 3D overhead crane working scenes and working conditions. Trained person operates the overhead crane in 3D visualization of the scene using a large screen projector or a stereo imaging lens. Virtual reality system can be helpful for the operator to match real-time interaction in the three-dimensional environment, and they can operate trolley, hook and all kinds of safety devices to acquire knowledge of the operating procedure. Virtual environment gives the opportunity to operators for repetitive training. Accident reconstruction and analysis are possible for identification of errors occurred in training process.

2.3 Virtual Reality Application in Training

Application of virtual reality nowadays makes it possible to reconstruct, simulate and visualize the whole operation process and find out the deficiencies that are present in the operation. It is being widely applicable in chemical, mining, construction, health care industries but its applicability in manufacturing industries is still an issue to be taken care of. Accident prevention training and operational safety programmes will be beneficial for a worker in a theoretical sense but practical approach to training through immersion and visualization is possible through virtual reality. Visualization of accident that had happened in the past and providing training to the employees in the virtual environment which will give them the feel of real environment to avoid those in future is to be added in safety training programmes. Virtual reality platform can emulate immersive interactive scenarios which reflect real-life situations, simulating the way equipment responds, replicating soft skills such as human actions and behaviour. The simulation of unpredicted control situations can reduce cost, complexity and time associated with the understanding and knowledge acquisition process needed for the real-life problem solution.

2.4 Advantages of the VR Training System

Compared with previous training and assessment methods, the overhead crane operating system based on virtual reality technology will have following advantages:

i. High Security: Using the overhead crane simulation system, one can safely carry out high-speed, extreme driving and very dangerous safety experiments, which cannot be achieved in the day-to-day experiment.
ii. Good reproducibility: Because the overhead crane is used in the complex environment in reality, the real-time experiment is hard to perform which can be easily achieved by VR technique, and the overhead crane simulator based on virtual reality technology can conveniently carry out data acquisition, model selection and setting of the simulation environment.
iii. Economic: Compared with the real vehicle test, the simulator occupies a small area and consumes less energy, and it can also be more convenient to set up various experimental conditions and experimental parameters in the software environment.

2.5 An Introduction to Simulators

Machine simulators are now widely used for giving training to the operators. In industry, education, military, the application of these simulators are increasing day by day. The current increasing trend of virtual reality is making that application more affordable. Virtual reality-based simulators are now used in various fields, and it is giving high return of interest. Virtual crane simulator is described below.

2.5.1 Virtual Crane Simulator

2.5.1.1 Environment and Needs

Workplace having cranes is very complex scenario. Workers should be properly trained and experienced to cope up with this complex environment. It is sometimes very difficult to reproduce the exact training scenario in real situations. Training process in real scenario is also hazardous in nature. Trainee may not be allowed to do any mistakes at the time of training in the real scenario but repetitive training can be accommodated in virtual reality training.

2.5.1.2 The Crane Simulator

The crane simulator is in general a virtual crane in computer-made 3D environment. The environment is the replication of a real environment. When operator or user wears Head Mounted Display, he/she will be immersed in that computer's simulated environment. He/she will feel that they are in that environment. They can walk through the environment and interact with the objects. The realism of the environment means not only the photorealistic view of the environment, but the rigid body physics, dynamics, kinematics, gravity, collision, elasticity also. All the intractable factors must be incorporated in the environment. Users can practice as much time as they want and make mistakes there. By making mistake they will understand the path which leads to accident. They will also understand about the barriers that must be incorporated to get rid of from those accidents.

2.6 Methodology for Development of Crane Simulator

The methodology considered here for developing virtual reality-based crane simulator model is mentioned step by step. Some prerequisites for building virtual reality-based scenario are learning of 3D modelling software, texturing, programming language (C ++, C#, python) and game engine. The methodological flow work is given in Fig. 2.1 for development of virtual reality-based prototype for crane operation.

2.6.1 3D Modelling and Texturing

3D modelling is the description of spatial vector location of objects, environments through a computer system. This section provides the basic concepts of the modelling technique. The method of using basic building block that is used for creating objects and the process of moving, rotating and resizing the objects are also explored here. The fundamental steps for 3D modelling are described below (retrieved from https://www.unrealengine.com).

2.6.1.1 Coordinate Systems

There are three perpendicular axes, which are labelled as the x, y and z axes in the field of 3D design. The origin is called the world origin that is used as the reference

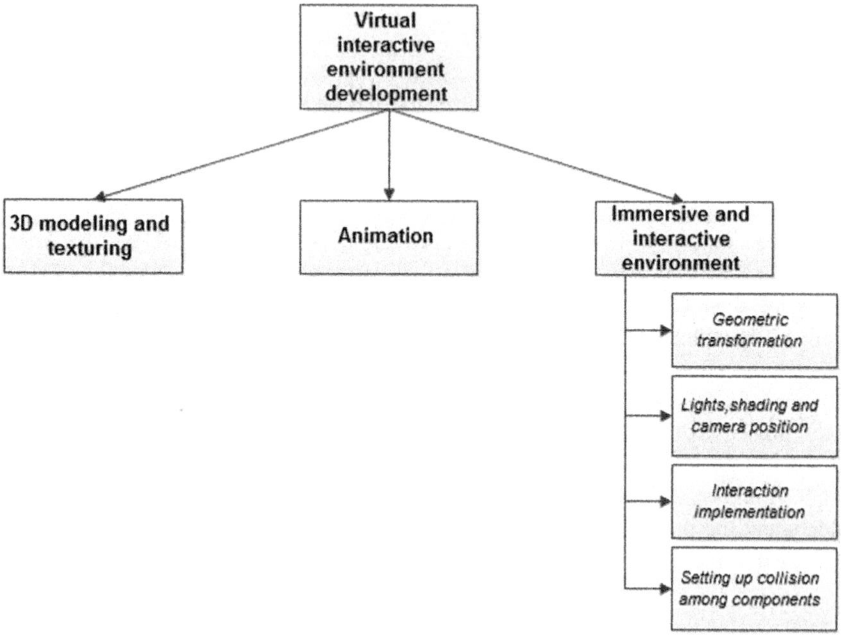

Fig. 2.1 Flow diagram of method

in 3D modelling space, and the three axes intersect here. Distance of a point in the space from the origin has been described by the x, y and z values. Rotation is also possible about the three axes, and these rotations are often referred to as pitch (rotation about the x axis), yaw (rotation about the y axis) and roll (rotation about the z axis). This coordinate system is named as the Cartesian coordinate system.

2.6.1.2 Points, Lines and Faces

The basic elements in 3D modelling are point, face and line also known as vertices, polygons and splines, respectively. A point is presented by the values along x, y and z axes, a line is also described defined by the XYZ values of its two end points, a face can be similarly defined by the XYZ values of the points that cover it (the points on a face must all be planar). Figure 2.2 shows an example of 3D models with coordinates.

Lines are made using points; multiple lines together make a face, and number of faces is combined to produce a shape or model. Some complex models, such as a character, may require millions of faces, while a simple box can be produced with just six faces. So it can be inferred in a common way that 3D objects are nothing but some combination of vertices. For reducing simulation time, least number of vertices in a model should be used.

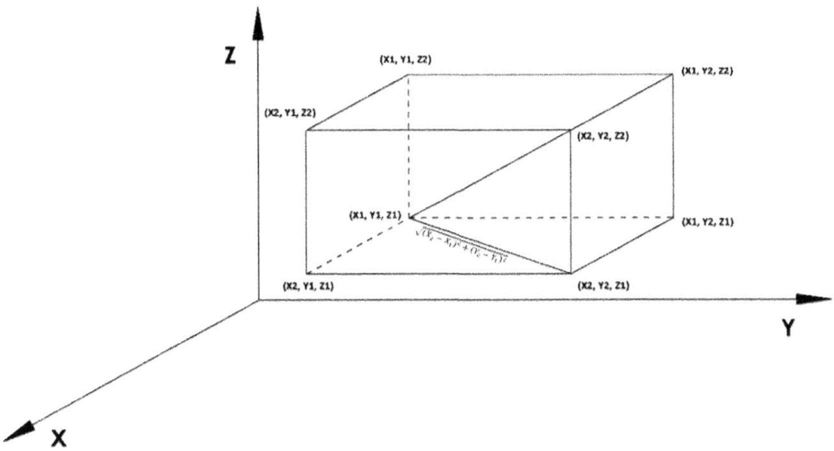

Fig. 2.2 Vertex, line, face

2.6.1.3 Image Mapping (Texture Mapping)

The 3D model above explained is only a mesh. Now to make that mesh look real, texturing is very much important. Texture mapping is a method for giving detailed surface texture (a bitmap or raster image) or colour to the 3D mesh. A shape or polygon gets live after applying texture map on their surfaces. In a simple way, texturing can be said as pasting some colourful pattern on white box. Every vertex of a polygon is a texture coordinate (in the 2D model, these vertexes are known as a UV coordinate). Figures 2.3 and 2.4 describe the essence of image mapping.

Fig. 2.3 3D model without texture

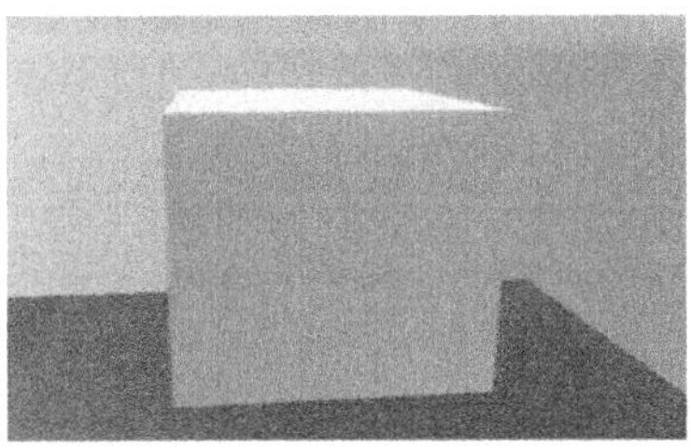

Fig. 2.4 3D model with texture

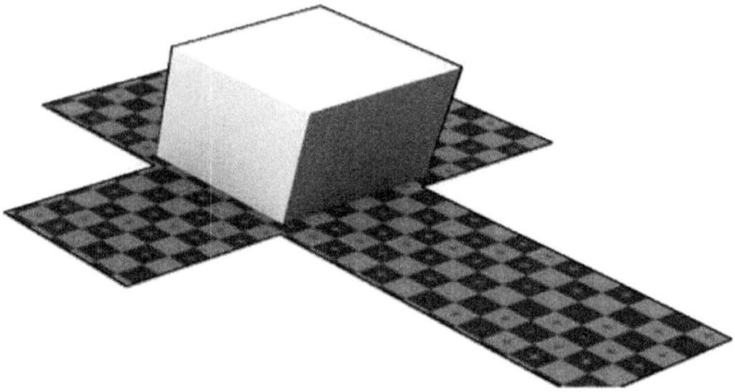

Fig. 2.5 UV mapping

Blender software is used for texturing. The following are the essential things that are very much useful for proper texturing.

UV Mapping

The process of projecting 2D images on the surface of 3D mesh. Figure 2.5 provides the UV map of a cube.

Normal Mapping

Normal map is used to provide the direction of normal of a geometric shape. Following gives the normal maps of cube (Fig. 2.6).

Bump Mapping

It is provided for visualization of irregular displacements on the surfaces of plane mesh.

Fig. 2.6 Normal mapping

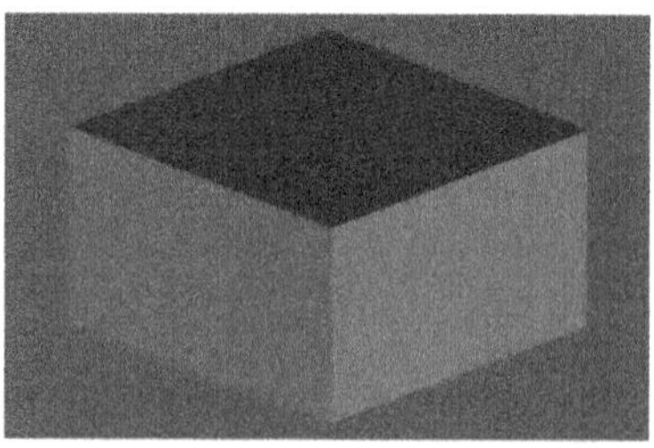

Reflection Mapping

This is needed to give the reflection of a 3D mesh in the environments.

After the texturing, the texture model has to be imported in virtual reality engine that is used for creating complex virtual scenario. Unreal engine or unity 3D can be used as per user's requirement.

2.6.2 Animation

Animation is to give life to a 3D model. The 3D static mesh, after animation, behaves like real character. Blend Space is one of the concepts that is widely used for animation. In Blend Space, different types of animation can be added and made a single animation. That makes the model more real. Suppose in one animation, the 3D avatar will walk and in another animation it will run. Now in Blend Space, these two animations can be grouped together.

2.6.3 Immersive and Interactive Environment

For making the virtual environment, geometric transformation, lighting and shadows and texturing have to be implemented for making immersive environment.

2.6.3.1 Geometric Transformations

Geometric transformations are used for modifying the size, shape of model proportion to other model and also used for the spatial position of the object in the world. The most common transformations that are used for this are translation, rotation and scaling. Transformations can be applied to entire world, single objects, or single primitives (points, lines and faces). The value of the axis should be specified in which the transformation is to be applied and which origin in the world has to be used at the time of transformation.

2.6.3.2 Lights, Cameras and Shading and Surface Characteristics

Overview of virtual lights and camera used for modelling technique and also the description of shading and surface options available in modelling software are given below.

Lighting and Shadows

Lights are used for illumination in 3D environments for focusing objects and are very important for the perfect realism of final virtual scene. When there is no light, the objects will looks like the following.

There are various types of light sources in 3D software; each of it gives different atmospheric situations. One can also control intensity and colour of light by manipulating it and give a different scenario according to user's specification. Following are various types of lighting sources that are used in this study.

Directional lights. The directional light provides parallel rays. Distant light sources such as sun where lights are parallel and coherent are simulated using this light. Following figure gives the overview of directional light

Point light. The point light rays come from a single point, outwards in all direction. It simulates various light sources such as candle, bulb. Following is the figure of point lights.

Spot lights. A spot light looks like a cone that transmits its light in a specific direction. This is modified by using the cone angle.

Figure 2.7 shows all the different lights and its visualization in the environment.

2.6.4 Interaction Implementation

Dynamic components in the environments are identified, and each are to be given the interaction effect with the help of blueprint or visual scripting in UE4. In UE4 game engine, blueprints are used to define object-oriented classes and add functionality to them like other scripting languages. Blueprints are created inside of UE4 editor visually, instead of by typing code, and saved as assets in a content package. These essentially define a new class or type of actor. Blueprints do not always need to contain scripted behaviour. Using the components present in the environment, blueprint can be made reusable to accelerate your level design process through event graph which is otherwise called as level blueprint. Modification in the class blueprint will have impact on complete project package. Event graphs contain the design-time and game-time behaviour of your blueprints. The event graph of a blueprint contains a node graph that uses events and function calls to perform actions in response to gameplay events associated with the blueprint. This is used to add functionality that is common to all instances of a blueprint. This is where interactivity and dynamic responses are set up. For adding functionality, model must be rigged to make it visually real.

2.6.4.1 Rigging

Rigging is like adding some skeleton in the static mesh, so that the model can move or can be deformed. Before animation, the model should be rigged otherwise proper

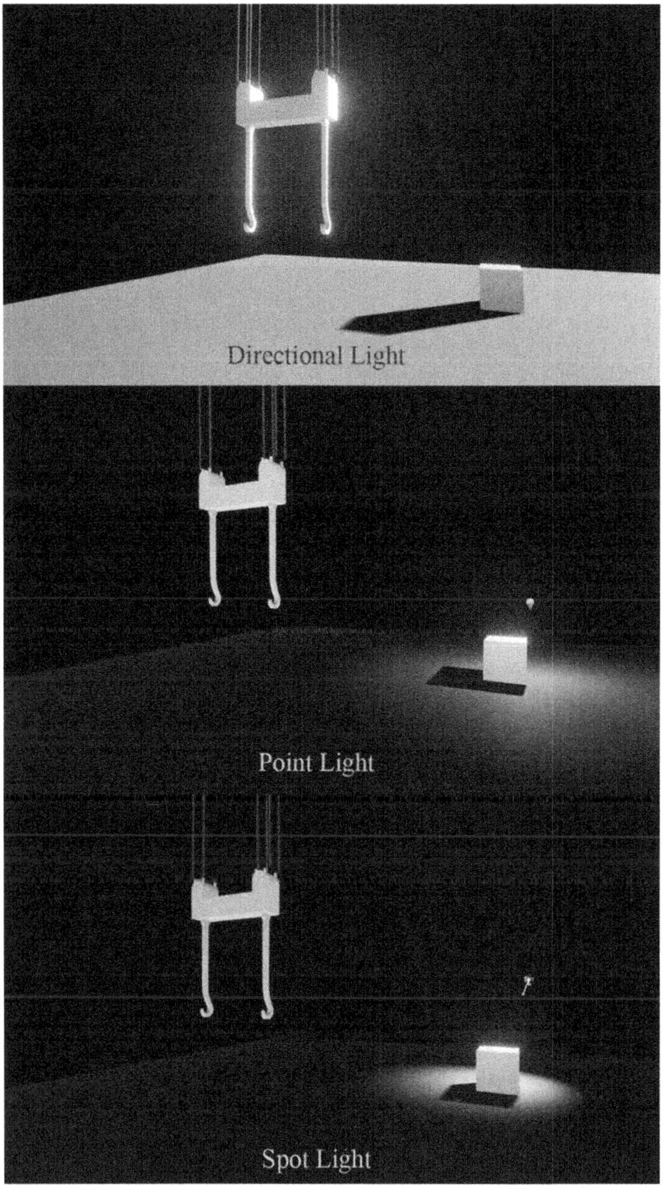

Fig. 2.7 Different light sources of the environment

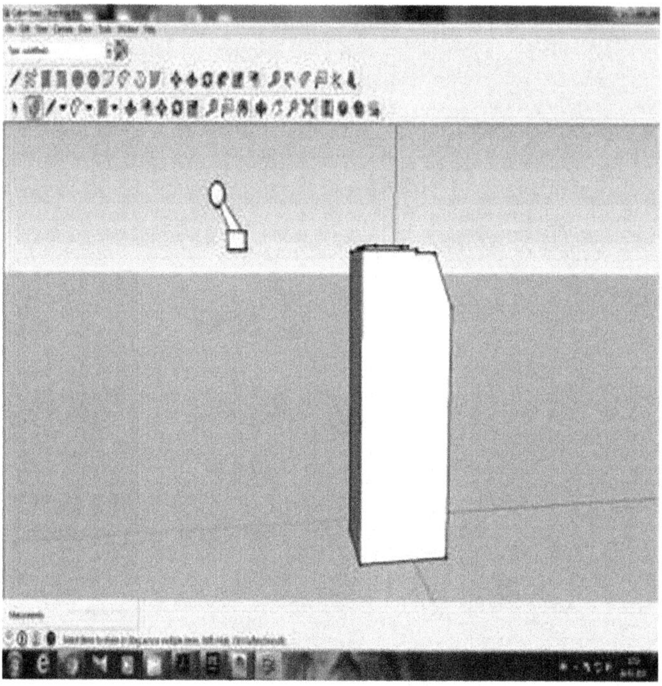

Fig. 2.8 Liver

animation cannot be availed. Figures 2.8, 2.9 and 2.10 are some examples of the process of rigging of the 3D model (http://www.autodesk.in/products/maya).

2.6.5 Setting up Collision Among Components

In virtual environment, to avoid user's walk through the static/dynamic models, setting up collision is important. Different types of collision detection techniques are available in UE4 game engine. Box collision technique is used in video games and other simulator, but here the collision is not proper in visualization, and computational time is more. Spherical type of collision is modified version of box collision where computational time is lesser than box collision but improper collision is still there. Convex hull collision is the most developed version of collision detection. Here collision is given according to convex hull algorithm that will detect all the boundary point of the model. Computation time of this algorithm is very high but still it is used everywhere due its efficiency.

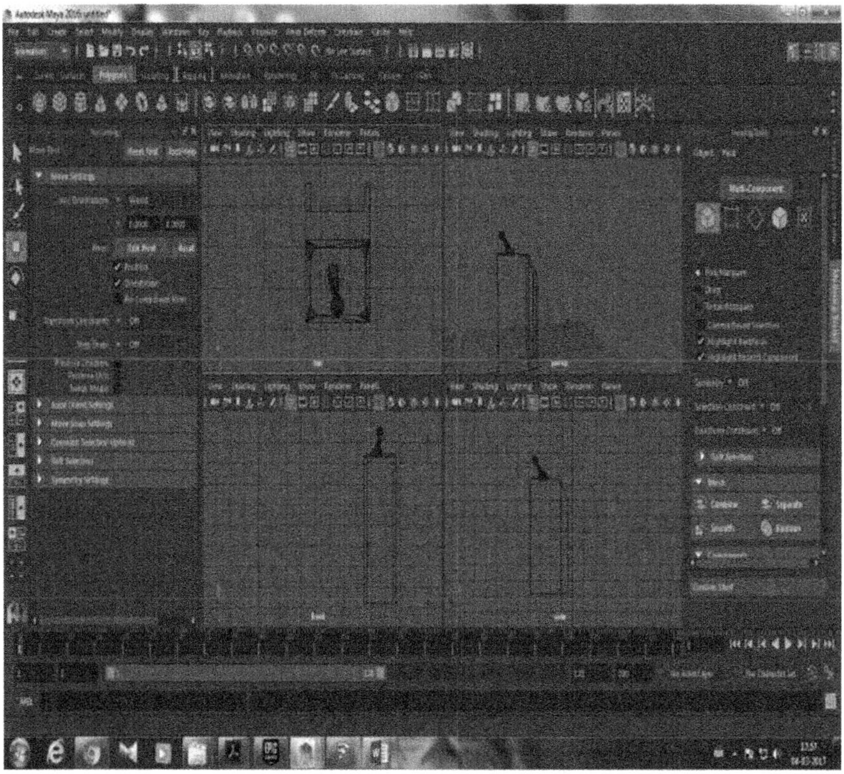

Fig. 2.9 Four views of liver

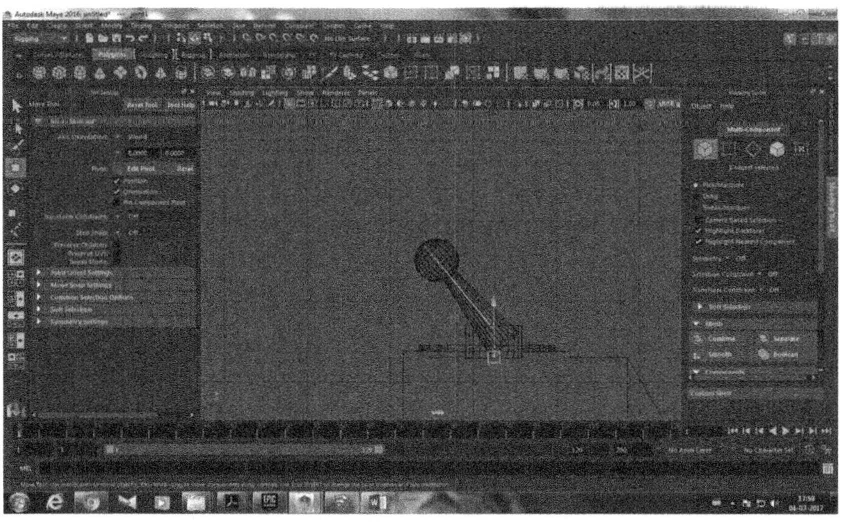

Fig. 2.10 Skeleton added to the liver

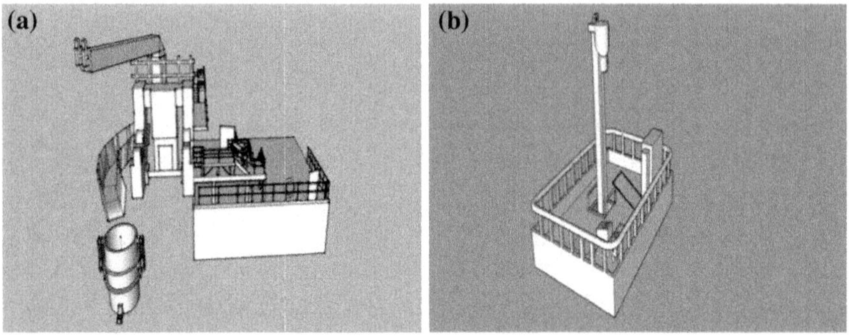

Fig. 2.11 **a** Caster, **b** nozzle

2.7 Results

Three-dimensional modelling is done using GOOGLE Sketch up Pro. EOT crane operation simulation is the focus of the current work for which pictures, and video data are collected from a steel industry where EOT crane operation is needed for steel-making process (https://www.sketchup.com). Glimpses of component modelling are shown in Fig. 2.11.

2.7.1 Textured Models

For giving custom texture to the model, those must be segmented as per its different sections. Model segmentation is done in Blender software. Those sections are then given different colour for understanding that this section is different from the other section. After the segmentation of the model, relevant images have to be collected for giving texture. Then the collected image has to be converted in normal map, occlusion map and specular map. Those map are then integrated with the original image using visual scripting language. We can change various property of the images using unreal engine software. The projection view can be taken care of using normal map. Reflection map can be used to give shininess of the model. Occlusion map can then be integrated for giving the property where indirect light falls on the model. Figures 2.12 and 2.13 show the actual mesh and textured mesh of caster and slag-spot stand.

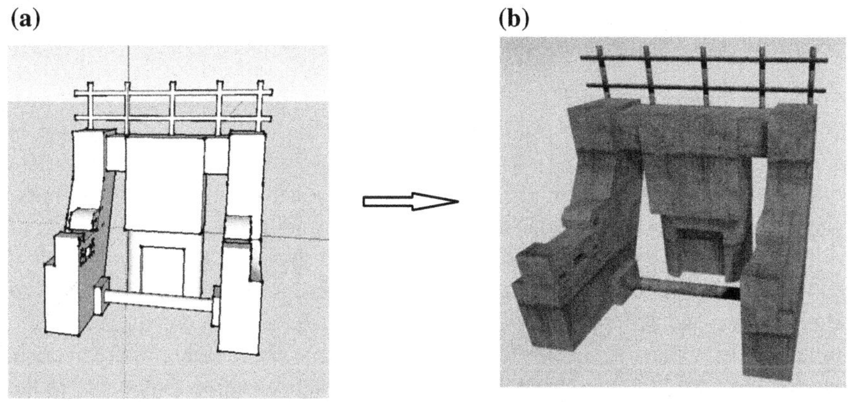

Fig. 2.12 **a** Caster (without texture), **b** caster (with texture)

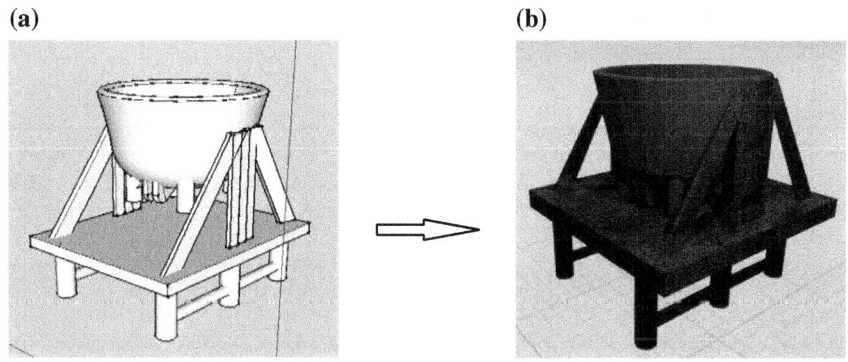

Fig. 2.13 **a** Slag pot (without texture), **b** slag pot (with texture)

2.7.2 Developed Virtual Environment

The developed environment can be simulated using keyboard and mouse of the computer. The benefits of this crane simulator is operator can be trained using virtual simulator and make mistakes as much as possible that is not possible in real scenario. Also for providing training to operator in real scenario whole operational process must be stopped to avoid mishaps. Stopping the operation means a huge loss to the organization. But using this simulator, operator can be trained without stopping the whole operation as this training is carried out remotely in digital computer. The simulator can be enriched by using various sensors technology with this virtual environment. Specific operations in the developed environment can be simulated using sensor-integrated hand data gloves. And using haptic technology, we can also the feel the touch of each operations carried out in the virtual environment. We can mount Head Mounted Display for being immersed in the environment.

2.8 Conclusions

Dangerous, impractical and expensive scenarios in real world are achievable nowadays owing to the courtesy of virtual reality. Due to more and more use of these VR applications, its cost is going down and it is becoming mainstream where we can expect its further applications in occupational safety at manufacturing industries. Application of virtual reality technique in modelling and simulation of overhead crane enabled trainees to perform planned and repetitive operations in a risk-free environment. VR training will enhance the operator's acquaintance with the workplace and different hazards present. It will be a boost for the operator's confidence, and the developed simulator will reduce the training cost of crane operators. Hazard identification and prevention strategies can be thought of in the further work.

Acknowledgements The authors gratefully acknowledge the management of steel plant for their support and cooperation.

References

Choi, S., Jung, K., Noh, S.D.: Virtual reality applications in manufacturing industries: past research, present findings, and future directions. Concurrent Eng. **23**(1), 40–63 (2015)

de Gennaro, S.: "The Virtual Crane." Accessed: 12 Dec 2016

Goulding, J., Nadim, W., Petridis, P., Alshawi, M.: Construction industry offsite production: a virtual reality interactive training environment prototype. Adv. Eng. Inform. **26**(1), 103–116 (2012)

Guo, H.L., Li, H., Li, V.: VP-based safety management in large-scale construction projects: a conceptual framework. Autom. Constr. **34**, 16–24 (2013)

https://www.unrealengine.com

http://www.autodesk.in/products/maya

https://www.sketchup.com

"OSHA Safety Training Courses." Accessed 15 Dec 2016

Chapter 3
Reduction of Coal Dust Exposure Using Water Mist System

Kanjiyangat Vivek and H. Manikandan

Abstract Quantification of airborne dust exposure is an essential step in eliminating lung diseases caused by over exposure to dust. A large number of workers employed in coal boilers are potentially exposed to significant amount of airborne coal dust during work period. The objective of the study was to assess the efficiency of water mist system in reducing airborne coal dust concentration in the working atmosphere near a coal-fired boiler. A sample of 8 workers was selected, representing approximately 80% of the total workforce involved in the coal handling operations. Exposure monitoring was conducted for 480 min during regular working hours. A personal dust sampler was used to capture the airborne coal dust (PM5) before and after installation of water mist system. After quantification and comparison, it was observed that the water mist system successfully reduced up to 81% of respirable dust from the worker's breathing zone. Further, the respirable dust concentration was reduced from 7.17 to 1.26 mg/m^3 which is significantly below from the recommended occupational exposure limit (OEL) of 2 mg/m^3. The same technology may be used for many other applications to reduce airborne dust levels and protect worker's health.

Keywords Water mist system · Personal sampler · Coal dust exposure
Respirable dust reduction · PM5

K. Vivek (✉)
Department of Safety Health and Environment, Dr. Reddy's Laboratories,
Hyderabad, India
e-mail: iamvivekka@gmail.com

H. Manikandan
Department of Mechanical Engineering, Birla Institute of Technology
and Science Pilani, Pilani, India

J. Maiti and P.K. Ray (eds.), *Industrial Safety Management*,
Managing the Asian Century, https://doi.org/10.1007/978-981-10-6328-2_3

3.1 Introduction

The health of workers in many industries is at risk through exposure to inhalation of dust which may contain toxic metal, fibrogens like silica and/or other toxic organic or inorganic chemicals. Regulatory agencies, in industrialized countries, lay great emphasis on monitoring of such exposure. Also in India, the 1948 Factories Act as amended in 1987 for factory management (Section 7-A (e) and 41-F) to ensure that workers should not be at risk of ill-health as a result of the work they perform. An occupational health and industrial hygiene program allows you to respond effectively to workplace injuries and illnesses and to monitor potential health problems. A workplace hazard analysis is an essential first step that helps to determine the sources of potential exposure and quantification of risk involved which can arise during normal operation. Industrial hygiene program includes the development of corrective measures to control health hazards by either reducing or eliminating the exposure. These control procedures may include the substitution of harmful or toxic materials with less dangerous ones, changing of work processes to eliminate or minimize work exposure, installation of exhaust ventilation systems, good housekeeping (including appropriate waste disposal methods) and the provision of proper personal protective equipment (Barbara and Patricia 2001).

Regulations require regular monitoring of the working environment in a factory to ensure that the airborne chemicals and pollutants are within prescribed limit. Since a worker does not stay at the point of high concentration for all the 8 h of exposures, the most practical way to monitor the risk is through quantitative industrial hygiene study using 'personal sampler' carried on the exposed worker's body and sampling the air from his actual breathing zone. There are many different methods of taking air samples, but by far the most widely used and preferred is to connect a battery-operated pump to a filter medium. The pump should be capable of drawing air through the filter at a constant rate for a time more than 8 h, even in adverse conditions such as extreme cold. This criterion is based on the recommendations that samples should be taken on a personal basis for an 8 h time weighted average (TWA).

3.1.1 Dust as Health Issue for Workers

Dust is scientifically defined as "small, dry, solid particles projected into the air by natural forces, such as wind, volcanic eruption, and by mechanical or man-made processes such as crushing, grinding, milling, drilling, demolition, shovelling, conveying, screening, bagging and sweeping" (IUPAC 1990). Dust particles are usually in size range from about 1 to 100 μm in diameter, divided into three fractions (i) respirable (D50 = 4 μm size) (ii) thoracic (D50 = 10 μm size) and (iii) inhalable (D50 = 100 μm size). ACGIH® has a TLV of 2 mg/m^3 for the respirable dust fraction of coal dust containing less than 5% quartz. Coal workers' Pneumoconiosis (CWP), commonly called Black Lung, can occur in workers working in coal boilers after more than 15 years of excessive inhalation of respirable coal dust. Lung parenchyma, lymph nodes and hilum can be affected due to chronic exposure to coal dust, and the onset of the disease is directly related to the airborne concentration of coal dust in the working environment and the total duration of exposure. In the present scenario, the exposure reduction is based on administrative control and the use of respiratory protection during specific work tasks (Von et al. 2010). Thus far, non-engineering intervention studies concerning airborne dust exposures have mainly focused on education to promote respiratory health through an increased and proper use of respirators (Dressel et al. 2007, 2009; Donham et al. 2011; Kim et al. 2012; Jenkins et al. 2007). The need for effective exposure control strategies for these workers is well acknowledged, but efforts have been historically challenging mainly due to intermittent nature of the performed work (Jenkins et al. 2007; Reynolds et al. 1996). The interest on direct exposure reduction through engineering control has been limited; also justified by few studies using quantitative measurements of exposure (Reynolds et al. 1996; Choudhary et al. 2012). Though use of respirators is considered a low-tier exposure control technique, an exception resorts only for cases where exposure control by other methods is not feasible or until other effective engineering and administrative methods are established (Rodes et al. 1991). The research work furnished in this paper evaluates the effectiveness of water mist system as an engineering control on reducing workplace respirable coal dust exposure during coal handling in a pulverized coal boiler environment.

3.2 Materials and Methods

3.2.1 Objective

A number of studies on particulate matter (PM) emissions from heavy industries (steel, cement, fertilizers, etc.), their chemical characterization and impact assessment have been conducted in the past but boiler has never been a topography for such investigations (Sharma and Pervez 2004a, b; Sharma and Pervez 2003a, b; Sharma and Pervez 2002). The specific objective was to assess the respirable coal dust, i.e. PM5 exposure before and after installation of water mist system in a boiler environment, and to ascertain the efficiency of engineering control over other methods of exposure control.

3.2.2 Design of Study

3.2.2.1 Sampling Sites

The site preferred for the study was the working premises of a pulverized coal boiler. A boiler is a closed vessel which transfers thermal energy to water or other fluids. The thermal energy for a boiler can be produced from combustion of any of several fuels, such as wood, coal, oil or natural gas. Electric steam boilers and nuclear fission steam boilers are also being used for heating water or other fluids on large-scale industrial applications. A pulverized coal-fired boiler is an industrial boiler that generates thermal energy by burning pulverized coal also known as coal dust. The major activities in a boiler operation are unloading of coal, shovelling of coal, the collection of ash and PLC operations. About 80% of the population engaged in unloading and charging operations were selected for the study. Water mist system was chosen as the engineering control, nominal cost and ease of installation favoured the selection. The subjects chosen for the study were exclusively engaged in various activities in boiler operations for more than 5 years.

3.2.2.2 Personal Sampler

The personal sampler is shown in Fig. 3.1, and Fig. 3.2 clearly depicts various parts of APM 801(Envirotech) which was used for the study. A cyclone head holding filter is to be worn on the collar/lapel to draw air from the breathing zone of the subject. The PM5 cyclone has been designed to cut off at 5μ (50% cut aerodynamic dia.-Inlet: 100 microns, Stage 1: 5 micron) as recommended by DGMS (Directorate General of Mines Safety, India). Air leaving the cyclone has particle size of 5 microns or less which accumulate on glass microfiber filter of 37 mm diameter housed in leak proof Teflon filter holder fitted at the bottom of the cyclone.

The sampler was mounted in the breathing zone of worker and allowed him to perform his work for 8 h. The accumulated coal dust on the filter was quantified

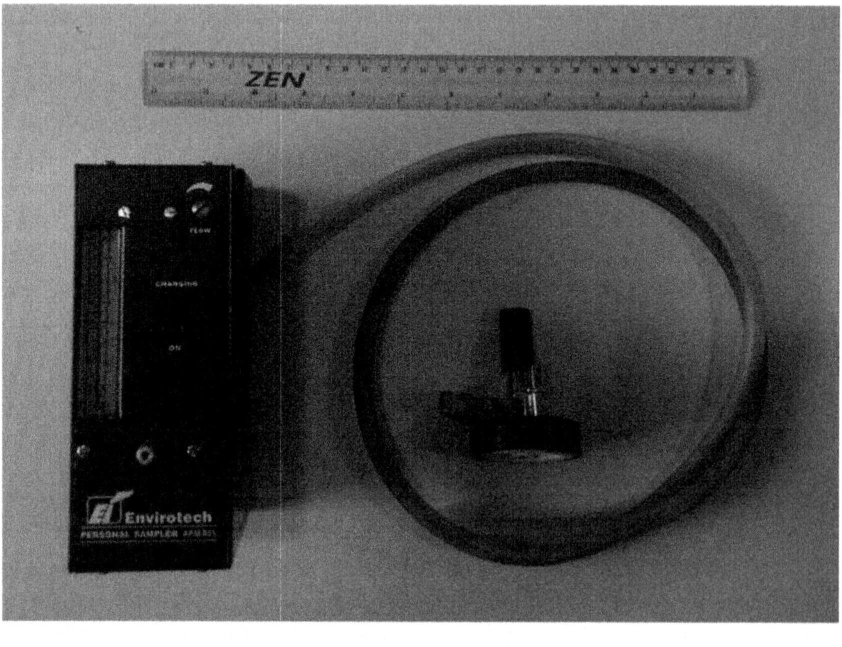

Fig. 3.1 Personal sampler (Envirotech-APM 801)

Fig. 3.2 PM5 cyclone

gravimetrically. To determine the mass concentration of the size fractions, the Particle Sampler filters were post-weighed on the same digital microbalance as was used to pre-weigh the filters. The calculation of the concentration of respirable dust was carried out through Data Spreadsheet (TSI, USA). For optimum efficiency of the cyclonic system, air flow rate was maintained at a rate of 3.11 L per minute.

3.2.2.3 PM5 Sampling

The eight subjects selected for PM5 samplings were adult male (25–55 y) engaged in unloading and charging of coal into the boiler. The identified subjects work for 48 h per week in the same environment. The personal sampling was conducted for eight hours (06:00 AM to 02:00 PM and 02:00 PM to 10:00 PM). Five samples were collected from each subject on alternate days excluding non-working days. Samples were collected on glass microfiber filters (Dia 37 mm) having a low resistance to airflow, a low affinity for moisture and a 98% collection efficiency for particles of 0.5 μ or large size (Whatman™: 1820-037). The glass microfiber filters used for the study possess fine particle retention and high flow rate as well as good loading capacity which are commonly recommended for gravimetric determination of airborne particulates stack sampling and absorption methods of air pollution monitoring.

3.2.2.4 Water Mist System

The mist system operates at 1000 PSI generating 10-micron-sized water droplets 80 number of nozzles were vertically mounted and spaced 0.5 m apart at the height of 3 m. The water mist system was continuously functioning during sampling. The water mist system increases the overall humidity at the study site, but the comfort level of the subjects was not subsided.

3.2.2.5 Statistical Analysis

Exposure data are presented as median, geometric mean (GM), arithmetic mean (AM) and range. The simple t-test for paired samples was used to evaluate differences between respirable coal dust concentration before and after installation of water mist system. The statistical analyses were done using the IBM SPSS software package of version 22 (Tables 3.1, 3.2 and 3.3).

Table 3.1 Respirable coal dust concentration (mg/m^3) before installation of engineering control

	N*	AM†	GM‡	Median	GSD§	Range
Subject-1	5	6.28	6.27	6.12	0.41	5.88–7.03
Subject-2	5	6.66	6.64	6.62	0.57	5.81–7.44
Subject-3	5	6.94	6.85	6.51	1.16	5.73–9.03
Subject-4	5	6.14	6.11	5.97	0.63	5.42–6.94
Subject-5	5	7.22	7.17	7.35	0.83	5.80–8.07
Subject-6	5	6.40	6.31	6.58	1.01	4.78–7.84
Subject-7	5	6.73	6.69	6.81	0.79	5.42–7.89
Subject-8	5	5.91	5.83	6.25	0.93	4.28–6.81

Table 3.2 Respirable coal dust concentration (mg/m^3) after installation of engineering control

	N*	AM†	GM‡	Median	GSD§	Range
Subject-1	5	1.70	1.64	1.95	0.43	1.04–2.19
Subject-2	5	1.68	1.63	1.79	0.38	1.10–2.23
Subject-3	5	1.44	1.38	1.49	0.42	0.98–2.08
Subject-4	5	1.70	1.70	1.67	0.15	1.50–1.97
Subject-5	5	1.63	1.60	1.76	0.30	1.11–1.93
Subject-6	5	1.59	1.57	1.69	0.29	1.23–1.97
Subject-7	5	1.35	1.26	1.39	0.47	0.77–2.08
Subject-8	5	1.61	1.59	1.60	0.22	1.27–1.93

Note All readings are taken between December 2015 and June 2016
*Number of measurements
†Arithmetic mean
‡Geometric mean
§Geometric standard deviation

Table 3.3 Comparison of respirable coal dust concentration (mg/m^3) before and after installation of water mist system

	Concentration before water mist system	Concentration after water mist system
Mean	6.53585765	1.588017891
Variance	0.864320318	0.140539802
Observations	40	40
Hypothesized mean difference	0	
Df	51	
t Stat	31.21711875	
P(T <= t) one-tail	3.30928E-35	
t Critical one-tail	1.67528495	
P(T <= t) two-tail	6.61856E-35	
t Critical two-tail	2.00758377	

3.3 Results and Discussions

A total of 8 subjects were participated in the study, and they had no difference in their work pattern. A total of 80 samples were collected from the selected subjects, 40 samples each was collected before and after the installation of engineering control. Measured respirable dust concentrations varied significantly after the installation of water mist system at the workplace. The two-tailed p-value is less than 0.0001, by conventional criteria; this difference is considered to be statistically significant. A comparison of respirable dust concentration before and after modification at coal handling area is represented graphically in Fig. 3.3. In the figure,

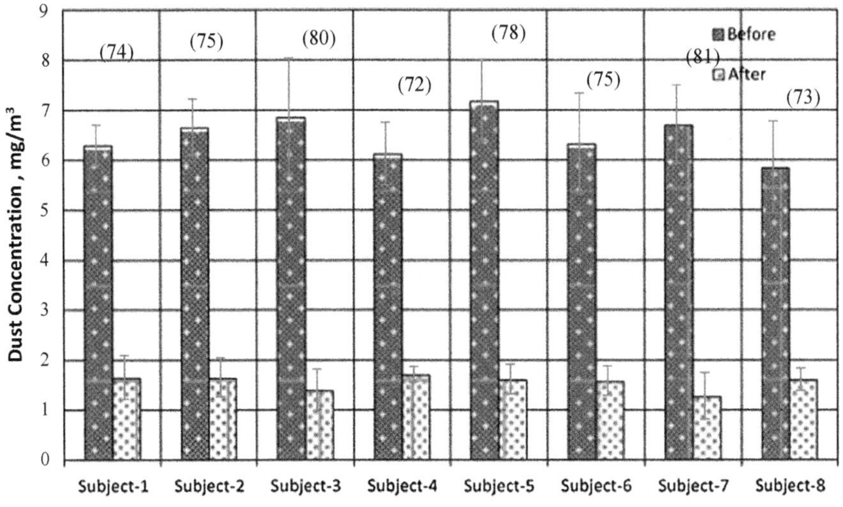

Fig. 3.3 Comparison of dust concentration before and after engineering control on each subject

values in parenthesis above the bar graph indicates the percentage reduction in dust concentration due to modification. It is revealed that there is significant reduction in total dust concentration 72 to 81%. The total dust concentration after modification varied from 1.26 to 1.70 mg/m^3 compared to that of 5.83–7.17 mg/m^3 before modification which is substantially lower than ACGIH TLV limits. The result of the modification is summarized in Table 3.2.

3.4 Conclusions

The statistical analysis of the data confirms a significant reduction in airborne coal dust and was observed from the boiler environment after the installation of water mist system. This water mist system reduced the respirable coal dust concentration in the range of 72–81% at the coal feeding section of the boiler. The respirable dust concentration was reduced from 7.17 mg/m^3 to 1.26 mg/m^3 after modification. The respirable coal dust concentration has reduced lower than recommended limit of 2 mg/m^3 at all the results post installation of water mist system. The developed system is very much feasible and cost-effective for all traditional coal boiler environment. The foundations laid by this research work can be used as a base for further studies in future to find out the effectiveness of each element in the hierarchy of exposure control. Further studies considering personal behaviour and work style are needed to validate the findings.

References

Choudhry, A.H., Reynolds, S.J., Mehaffy, J., et al.: Evaluation of parlor cleaning as an intervention for decreased occupational exposure to dust and endotoxin among dairy parlor workers - a pilot study. J. Occup. Environ. Hyg. **9**, 136–140 (2012)

Donham, K.J., Lange, J.L., Kline, A., et al.: Prevention of occupational respiratory symptoms among certified safe farm intervention participants. J. Agromedicine **16**, 40–51 (2011)

Dressel, H., Gross, C., de la Motte, D., et al.: Educational intervention decreases exhaled nitric oxide in farmers with occupational asthma. Eur. Respir. J. **30**, 545–548 (2007)

Dressel, H., Gross, C., de la Motte, D., et al.: Educational intervention in farmers with occupational asthma: long-term effect on exhaled nitric oxide. J. Investig. Allergol. Clin. Immunol. **19**, 49–53 (2009)

IUPAC: Glossary of atmospheric chemistry terms, International Union of Pure and Applied Chemistry, Applied Chemistry Division, Commission on Atmospheric Chemistry. Pure Appl. Chem. **62**(11), 2167–2219 (1990)

Jenkins, P.L., Stack, S.G., Earle-Richardson, G.B., et al.: Screening events to reduce farmers' hazardous exposures. J. Agric. Saf. Health **13**, 57–64 (2007)

Kim, J., Arrandale, V.H., Kudla, I., et al.: Educational intervention among farmers in a community health care setting. Occup. Med. (Lond) **62**, 458–461 (2012)

Plog, B.A., Quinlan, P.J.: Fundamentals of Industrial Hygiene: National Safety Council (2001)

Reynolds, S.J., Donham, K.J., Whitten, P., et al.: Longitudinal evaluation of dose-response relationships for environmental exposures and pulmonary function in swine production workers. Am. J. Ind. Med. **29**, 33–40 (1996)

Rodes C.E., Kamens R.M., Wiener R.W.: The significance and characteristics of the personal activity cloud on exposure assessment measurements for indoor contaminants. Indoor Air **2**, 123–145 (1991)

Sharma, R.K., Pervez, S.: Spatial variability of ambient particulate matter around a phosphatic fertilizer plant in India. J. Sci. Ind. Res. **61**, 1077–1083 (2002)

Sharma, R.K., Pervez, S.: Seasonal variation of PM10 and SPM levels around a cement plant. J. Sci. Ind. Res. **62**, 827–833 (2003a)

Sharma, R.K., Pervez, S.: Spatial and seasonal variability of ambient concentration of particulate matter around an integrated steel plant: a case study. J. Sci. Ind. Res. **62**, 838–845 (2003b)

Sharma R.K., Pervez, S.: Characterization and enrichment of selected toxic elements in ambient particulate matter around phophatic fertilizer plan-A case study. J. Sci. Ind. Res. **63**, 949–956 (2004a)

Sharma R.K., Pervez S.: Chemical characterization and enrichment of selected element in ambient particulate matter emitted from the cement plant- A case study. J. Sci. Ind. Res. **63**, 376–382 (2004b)

Von Essen S., Moore G., Gibbs S. et al.: Respiratory issues in beef and pork Production: recommendations from an expert panel. J. Agromed. **15**, 216–225 (2010)

Chapter 4
A Study on Performance Parameters Associated with the Effectiveness of Antilock Braking System on Rough Roads

N. Vivekanandan and Ajay Fulambarkar

Abstract In today's world due to advancement in automotive technology, there is a tough competition prevailing in developing high-speed vehicles. The high-speed vehicle has to perform better in terms of comfort and safety during heavy braking even in the rough road conditions. With the introduction of antilock braking system (ABS), the stopping distance of vehicle is shortened and steerability is possible under heavy braking condition, thereby avoiding collision. But the performance of ABS deteriorates in rough road conditions, where continuous contact between the wheel and the road surface is lost. There are number of parameters associated with the performance of ABS in rough roads such as fluctuation in wheel speeds, tyre parameters and type of suspension system. This paper deals with the study of various parameters and techniques that influence the performance of ABS in rough road condition. This study suggests a system that will improve the performance of ABS even in rough road conditions.

Keywords Antilock braking system (ABS) · Rough road · Algorithm
Wheel speed

N. Vivekanandan (✉)
Department of Mechanical Engineering, D. Y. Patil Institute of Engineering and Technology, Pune, India
e-mail: n.vivekanandan@pccoepune.org

N. Vivekanandan
Department of Mechanical Engineering, Pimpri Chinchwad College of Engineering, Pune, India

A. Fulambarkar
Pimpri Chinchwad College of Engineering, Pune, India

J. Maiti and P.K. Ray (eds.), *Industrial Safety Management*,
Managing the Asian Century, https://doi.org/10.1007/978-981-10-6328-2_4

4.1 Introduction

The demand for Sport Utility Vehicles (SUVs) is increasing year by year. According to JATO (2016) research report globally in the first quarter of 2016, the market share of SUV segment jumped to 27.4% compared to 23% in the same period of 2015. SUV growth also drove premium car volumes, with 2.15 million units sold, 833,300 of which were SUVs. According to the Global Market Research Company, J.D. Power, the growth and demand of SUV will be reached to top-position in the market among the passenger vehicles by 2020. Generally, these SUV's are used in off-road conditions. The major challenge associated with the SUVs is the implementation of effective braking system. With the advancement in automotive technology, lots of improvement in braking system is visible and with the introduction of antilock braking system (ABS) in modern vehicles, stopping distance after applying brakes have reduced drastically thereby increasing the safety of occupants. But the effectiveness of ABS deteriorates in rough road conditions and major challenge is the SUVs are mainly designed for rough road conditions. Hence, improving the braking condition of vehicle in rough roads also will provide safety to the drivers travelling on those roads.

Rough road excitations are unpredictable and only less works have been carried out to increase the effectiveness of ABS on rough roads. There are number of contributing factors that affects the performance of ABS on rough roads such as effect of suspension on ABS, effect of tyre oscillation, vehicle body motions such as roll, pitch and yaw that are excited by rough road input, Run-in effect of tyre force generation, sudden changes in wheel speed as the wheel encounters an obstacle, controller malfunction, Tyre Model considered while designing, modelling of test vehicle, Road Condition modelling. Literatures are available for increasing the performance of ABS considering each parameter separately or few parameters in combinations but the outcomes were not much satisfactory. This paper deals with the study of all the performance parameters associated with the increasing the effectiveness of ABS on rough road condition as highlighted in different literatures. Further investigation will be carried out to consider all the parameters influencing the performance of ABS to develop a more effective and reliable braking system for rough road condition.

4.2 Principle of Antilock Braking System

Under panic braking,the wheels lock and the vehicle starts skidding. This condition leads to number of consequences such as increase in braking distance, loss of steering control and tyre wear. The main effect is that an accident could occur. A force generated by application of brake hampers the vehicle motion by applying a force in the opposite direction.

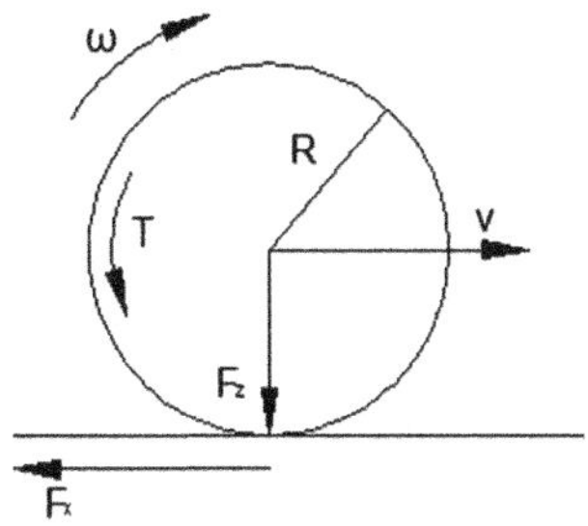

Fig. 4.1 Single wheel-car braking model

To simplify the problem, the single wheel-car braking model is shown in Fig. 4.1. where v is the speed of the vehicle; F_x is the braking force from the road; F_z is the vertical load of the wheel; R is the radius of the wheel; I is the rotary inertia of the wheel; ω is the angular velocity of the wheel; T is the braking torque of the brake; μ is the friction coefficient between wheel and road.

In severe braking conditions, a point is reached in which the tangential velocity of the tyre surface and the velocity on road surface are different where an optimal slip corresponding to the maximum friction is obtained. The wheel slip, λ is defined as:

$$\lambda = \frac{v - \omega R}{v} \tag{4.1}$$

where R, ω, and v are the wheel rolling radius, wheel angular velocity and the vehicle forward velocity, respectively. In normal driving conditions, $v = \omega R$, therefore $\lambda - 0$. Under panic braking, it is general to have wheel lockup at $\omega - 0$ while $\lambda = 1$. Wheel lockup extends the stopping distance and directional control is lost (Stan et al. 2007; Drakunov et al. 1995).

Figure 4.2 shows the relation between slip ratio λ and friction coefficient μ. ABS control principle is controlled by modulating wheel angular velocity through controlling braking torque T, so that the wheel slip rate λ is controlled at the best slip rate λ_0 and the friction coefficient keeps at near the μ_0. This device reduces the

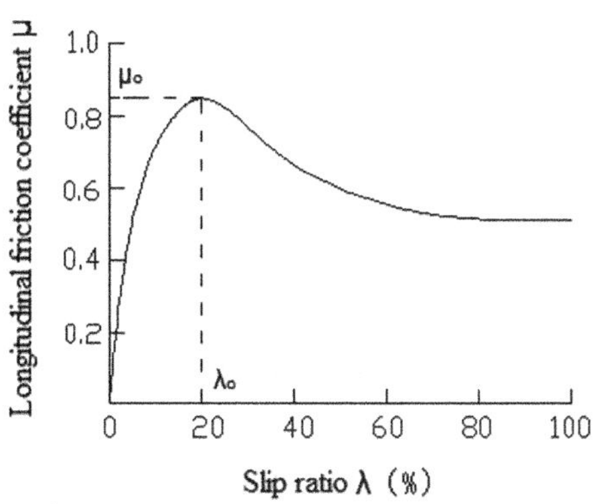

Fig. 4.2 Relation between slip ratio and road friction coefficient

braking distance and ensures the vehicle directional stability during braking. The main difficulty in the design of ABS control arises from the strong nonlinearity and uncertainty of the problem.

4.3 Performance Parameters Affecting the Effectiveness of ABS on Rough Road Condition

4.3.1 Effect of Suspension on ABS

On flat roads, the suspension setting has very less effect but on rough roads, the suspension setting plays a significant role in braking performance of vehicle with ABS. Few studies have been carried out earlier that used simplified quarter car model, pitch bounce model and simple modification of damper. Hamersma et al. (2013) suggested that the braking distance with ABS deteriorates up to 20 m compared to braking without ABS applied at 70 km/hr. It was concluded that by using different combination of spring and dampers the stopping distance can be reduced up to 14 m.

Herman et al. (2014) investigated the braking performance of the vehicle with ABS by using a semi-active suspension system on two different undulating road profiles such as Belgian paving and parallel corrugations. Modifications were made in 4S4 suspension unit developed by Thoresson (2007). The system developed had provision to control the stiffness and damping of front and rear suspension. Best results on both the roads were achieved with soft spring on the rear suspension and worst braking performance with soft springs on the front suspension.

Niemz and Winner (2006) used an approach called Mini Max control logic that changes suspension from soft to hard and vice versa. By this controllable body dynamics, the stopping distance reduced by 1.5% compared to best passive damping.

4.3.2 Effect of Tyre Oscillation

The rough road has a significant effect on the tyre oscillation. According to torsional dynamics of tyres, the wheel inertias response is coupled to ring inertia. The deflection between the wheel and ring depending on the stiffness, damping and mass distribution will cause significant error in the estimation of tyres longitudinal slip. Because of these errors introduced under aggressive braking, affects the ABS algorithm and reduces the braking performance.

Van der Jagt et al. (1989) suggested that axle oscillations because of rough road resulted in normal load variations and there by increases stopping distances and reduced lateral control of the vehicle. This is more significant at the resonance region of the suspension system.

Adcox et al. (2012) investigated the system developed considering the side wall flexibility, transient and hysteretic trend, ground friction effects and hydraulic braking system. They developed a system considering wheel acceleration-based ABS controller similar to a commercial ABS algorithm. The results strongly correlated the stopping distance and ABS control without having error in estimating the tyres longitudinal slip under tyre oscillations.

4.3.3 Vehicle Body Motion

Reul and Winner (2009) optimized the braking on rough road by integrating ABS and Continuous Damping Control (CDC). Generally, vehicle body motion such as roll, pitch, and yaw disturbs the slip control of the wheel. The theoretical approaches for deriving the strategies are used to deal with sharing the information between ABS and CDC in order to improve the slip control quality and adjusting braking torque or wheel load coordinately. Test has been performed considering dynamic wheel load (due to pitching and lifting) for calculation of braking force operation point. It was also found that the operating point changes more, if dynamic wheel load information is implemented in ABS control.

Penny and Els (2016) developed a modified Bosch algorithm and demonstrated that ABS performance deteriorates on rough terrains as a result of loss of wheel contact occurring due to severe roll and pitch motions with varying load transfer.

4.3.4 Run-in-Effect of Tyre Force Generation

As described by Gillespie (1992), there is a lag of one and half revolutions of wheel to reach steady-state force condition from start and is termed as run-in-effect and this affects the tyre force generation when operating in rough road conditions. When operating in rough road condition, varying vertical load on the tyre makes the contact patch generation to lag because of time taken by rubber to deform after application or removal of vertical load.

4.3.5 Sudden Changes in Wheel Speed, as the Wheel Encounters an Obstacle

Generally, four wheels ABS has disadvantage that the stopping distance extends on rough road and system complexity, size, weight, and cost. Satoh and Shiraishi

(1983) developed a new four wheel ABS that employs "select high" technique for the front wheels and "select low" technique for rear wheels. The select high technique also simultaneously controls braking torque for the right and left wheels, contrast with select low technique, in response to a signal from either of them which is predicted to get locked later. Therefore, it may permit the other wheel to get locked and reduce the stopping distance on rough road.

Jiang and Gao (2000) developed a cost effective method to estimate the vehicle velocity by using an adaptive nonlinear filter. The method is only based on wheel speed measurement without considering the factor of acceleration. But off-time testing result showed that an accurate and smooth estimation is achieved. As both the road conditions and decelerations are not known at the beginning, the error obtained is of not much importance when ABS is applied first. But, in actual demonstration, it is found that the proposed algorithm is able to recover from the initial error and converge to the actual vehicle velocity.

Daiss and Kiencke (1995) proposed a fuzzy-based system that uses a multi sensor data fusion integrated with commonly used wheel speed sensor to predict the vehicle speed with an accuracy of 1%. Multiple sensors are used to measure vehicle speed and an estimator decides which sensor is more reliable. Gyroscope is used to measure yaw rate, speed sensors are used to measure speed in each wheel, an accelerator sensor for longitudinal acceleration are used.

4.3.6 Controller Algorithm

The present ABS is controlled by slip ratio and wheel acceleration. In rough road due to large fluctuations in wheel velocity, slip ratio leads to poor performance of ABS.

Watanabe and Noguchi (1990) proposed a new algorithm that shifts the wheel acceleration threshold, which determines the initiation of the brake-pressure reduction by the road friction coefficient and the disturbance and by determining the proper duration for the pressure reduction by the road friction coefficient at same time. The proposed algorithm reduced the velocity fluctuation, thereby leading to effective steerability and shorter stopping distance.

Aly et al. (2011) reviewed the methods used in the design of ABS system highlighting the main difficulties and summarized the more recent development in their control techniques. Different ABS controls such as classical, optimal, non-linear, robust, adaptive and intelligent control are discussed.

Day and Roberts (2002) suggested that accurate simulation modelling of the vehicles fitted with ABS during braking as well as combined braking and steering manoeuvres requires the effect of ABS system on the resulting vehicle trajectory to be included. Simplified lump parameters are not adequate for detailed 3-Dimensional vehicle simulations.

Generally, in classical Bosch algorithm, the pressure is not decreased in step-wise manner in phase 3 as shown in Fig. 4.3.

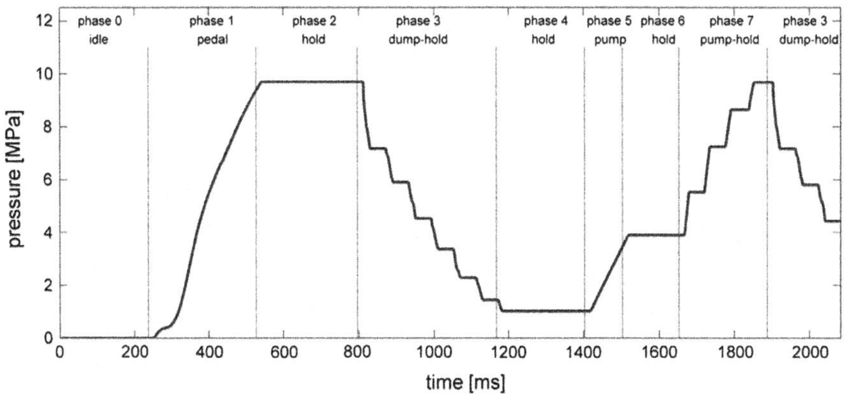

Fig. 4.3 Brake pressure, phases, and states for a typical control cycle on a high traction smooth surface. [Bauer, and Bosch (1999)]

To keep the longitudinal slip of tyre below the set threshold, it dumps the pressure as quick as possible in less time. But because of delay in the response of mechanical relays, the classical algorithm is ineffective as the controller senses too late that lateral stability has been regained and the pressure is lost.

Penny and Els (2016) suggested a modified algorithm to address the response delay of the electro hydraulic hardware. Further, an accelerometer is added to deactivate the "Yaw moment build up delay system" [GMA] at high lateral acceleration. GMA is generally beneficial in μ-split scenario, it responds to a brake in the turn scenario by increasing dynamic load at the wheels and results in over steering effect.

A lot of research has greatly improved the performance of the integrated framework by utilizing various algorithms such as back-stepping control design, sliding mode control, and fuzzy control. The basic assumption in all these algorithms is that the dynamic system is available and they cannot be employed to control integrated system if they are not dynamic. Hence, Wang et al. (2008) suggested a quarter car vehicle models integrated with ABS and active suspension system on the basis of sliding mode control theory SIMULINK software was used to verify the effectiveness of integrated controller in the normal driving condition. The integrated sliding mode controller showed good control performance. They used structural learning by fuzzy-neural networks (FNNs), Hierarchical T-S FNN to identify unknown integrated system with a potential to handle very large number of input variables and reduced computational time and fuzzy rules. The two unknown nonlinear systems are the ABS and the active suspension of a road vehicle.

Fargione et al. (2016) developed a system based on fuzzy controller proposed and braking vehicle performance is optimized by genetic algorithm. The system developed is based on both fuzzy and genetic algorithm. Fuzzy controller was designed on the basis of behaviour of the vehicle during braking and establishing the variables suitable for control. The control system was perfected by means of training in genetic algorithm. The fuzzy system has many advantages including robustness, whose results are quantitatively similar to the once reached by an expert driver.

4.3.7 Tyre Model

The choice of tyre model ultimately determines the success of any ABS simulation. The tyre model must be capable of accurately describing the lateral, longitudinal and vertical tyre force. But, generally in most vehicle dynamic simulations, tyres increase the complexity of the problem because they are highly nonlinear vis-coelastic and rough road adds to the challenge in mathematical modelling.

Ozdalyan and Blundell (1998) used an integration of an ABS suspension and tyre model in ADAMS, but they are applicable to flat road simulation only. Jaiswal et al. (2010) suggested that transient tyre characteristic can have significant effect in vehicle handling in ABS and it involves wheel speed oscillation as a result of rapid changes in wheel brake pressure. Fourteen-degree-of-freedom vehicle model in combination with different single-point contact tyre models such as stretched hyphen string-based model, a modified stretched string model, and contact mass model is considered. It was concluded that the contact mass model exhibited reduced oscillatory behaviour and fast response time to the brake-pressure variation.

Pacejka 2002 tyre model is the best of the semi-empirical model class, and some modified versions such as stretched string or constant mass model are developed models for ABS. The Short Wavelength Intermediate Frequency Tyre (SWIFT) model is often recommended for ABS simulation. This is a semi-empirical Pacejka 2002 model combined with rigid ring model. F-Tyre model is an advanced tyre model that uses a flexible ring connected to rigid hub and can accurately simulate driving on rough road and is accurate to high frequency excitations. F-Tyre is the best tyre model considering braking in rough terrain conditions.

4.3.8 Modelling of Test Vehicle

Hamersma et al. (2013) used a 15-degree-of-freedom full car model of the test platform including drivers, passengers, and outriggers modelled in ADAMS, and it is used with cosimulation in MATLAB/SIMULINK for braking on rough terrain.

Thoresson et al. (2009) proposed a methodology for the efficient determination of gradient information, for an off-road vehicle's suspension system. They modelled a recreational off-road vehicle in MSC ADAMS, and coupled it to MATLAB for the execution of the optimization. For the optimization of the spring and damper characteristics successive approximation method, Dynamic-Q was used along with the optimization of both ride comfort and handling. The objective function value was determined using numerical simulations. The models were validated against experimental results. The simplified vehicle models had less numerical noise than the full vehicle simulation model and consumed less computational time.

Hamersma and Els (2014) proposed a system considering the effects of spring and damper characteristics on the braking performance of a sports-utility-vehicle (SUV) on hard, rough terrain. The simulation-based approach was used, using an experimentally validated full vehicle model of the SUV, built in Adams in cosimulation with MATLAB and Simulink. The simulations were performed on measured road profiles of a Belgian paving and parallel corrugations. The results clearly indicated that the suspension system has a significant impact on the braking performance, resulting in differences in stopping distances of up to 9 m.

Botha (2011) suggested due to the nonlinear dynamics of the tyre road interface as well as those of the vehicle as a whole during high lateral accelerations a high speed cornering of an off-road vehicle poses considerable challenges to the development of an autonomous vehicle. Practically because of highly nonlinear behavior, the lateral acceleration increases, during high-speed manoeuvres. In this study, they presented two robust driver models for use in an autonomous vehicle capable of path following at both low and high. Both models made use of the relationship between the yaw acceleration and steering rate to control the yaw angle of the vehicle. The first driver model was derived from the simulation of a full nonlinear vehicle model in ADAMS using Magic Tyre Formula. The second driver model was a mathematical model which incorporates a form of sliding control using Pacejka 89 tyre model. To reduce the cross-track error both driver models are coupled with a gain scheduling proportional derivative controller. They implemented two driver models on a Land Rover Defender and experimentally validated by performing a double lane change manoeuvres at speeds up to 80 km/h. Even though the lateral accelerations experienced were 80% of the vehicle limits the vehicle remained stable. The controller was capable of path following at various speeds and at high lateral accelerations.

4.3.9 Road Condition Modelling

Belgian Paving at Gerotek Test Facilities (Gerotek Test Facilities 2015) is usually used in for road condition modelling for testing ABS in rough road conditions.

Becker and Els (2014) measured the three-dimensional Belgian Paving profile with great accuracy to be used as a road profile in simulation, and was found to approximate a class D road as per ISO 8608 (ISO 8606: Mechanical Vibration–Road Surface Profiles 1995). The simulation model is set to run at the same discrete rate as the embedded computer at 1000 Hz. A driver model is implemented to ensure that all simulations are done in a straight as possible line Botha (2011).

Becker and Els (2014) did a study on economic way of getting profiles of rough surface (Fig. 4.4) for vehicle dynamics simulations. Because of the severe roughness commercially obtainable inertial profilometers was not able to profile the rough surfaces. A mechanical profilometer was built up and estimated by profiling

Fig. 4.4 Can-can profilometer measuring the Belgian paving (Becker and Els. 2014)

obstacles with rough 3-D test track profiles and known profiles. A good correlation among the profiled and actual terrains was obtained. Realistic three-dimensional (3-D) surface models were created from the terrain profiles. The Displacement Spectral Densities (DSDs) of the profiled terrains was obtained to have discrete peaks; a straight line fit was not an accurate estimation for the specific rough surfaces. The roughness index of the terrains profiled with the mechanical profilometer is notably more than the terrains normally profiled by inertial profilometers. The aim of the study was to develop a spectral method to get the frequency response of half-vehicle under pavement roughness. The measurement of vertical pavement profile was carried out along two roads sides. The surface roughness was indicated as spectral density function. To get the angular and vertical modal vehicle dynamic response along with the excitation of the power spectral density (PSD) of the pavement roughness a frequency response analysis was used.

The results showed that the vehicle suspension mode was increased because of the unpaved track signature at low speed. First vehicle vibration mode at 120 km/h had a significant motion amplification that caused passenger discomfort in an undulated asphalted road.

4.3.10 ABS Hydraulics

Accurate modelling of ABS hydraulics is extremely challenging, therefore many researchers use Hardware-in-the-Loop (HIL). Heidrich et al. (2013) introduced the architecture and technical realization of HIL platform devised for development and testing of integrated vehicle control systems. The proposed HIL platform had a provision of testing various configurations of steering, brake systems, dynamic tyre pressure control.

Theron and Els (2007) considered the mathematical modelling of a suspension unit. The unit has a hydraulic cylinder, the un-sprung mass, two nitrogen-filled accumulator springs, and damper ports. The deflection rate is used as input and repeatedly uses fluid dynamics theory to find the flow rates from every accumulator to the cylinder. The pressure was estimated by time-integrating the flow rates in the accumulators by determining the gas volumes with ideal gas theory. The output is the dynamic force and model predictions were compared with measurements.

4.4 Scope for Further Work

An integrated controller for the antilock braking system (ABS) and the active suspension system can be developed and evaluated on the vehicle to reduce the stopping distance on rough roads. The controller for the integrated system should be able to handle a large number of input variables. Besides the conventional sliding mode control, there are several control schemes including adaptive, fuzzy logic and neural network control approaches, and hybrid controllers. The algorithms for the controller should be able to identify the condition of the road surface. More attention should be given to the accuracy of the tyre model. The Magic Formula approach may not be the ideal, especially on simulations based on an undulating road. But F-Tire may lend more credibility to the simulations as the torsional dynamics of the tyre on the braking performance can be more accurately captured by the F-Tire tyre model.

The system will be more efficient if the suspension modes are intelligently selected based on vehicle speed or wheel hop excitation. Using artificial intelligence techniques, one may teach the vehicle to identify the road it is driving on and accordingly adjust the suspension to ensure optimum braking performance.

To measure the accuracy of the calculated longitudinal slip, a more advanced method of estimating the velocity of the vehicle, like use of individual wheel-centre velocities as separate reference velocities together with a Kalman filter for velocity estimation, can be considered. An effective ABS for rough terrain, considering all these factors, will be developed and this literature survey will form the effective background for the successful development.

4.5 Conclusion

The relationship between the ABS parameters and components is highly complicated because of its nonlinearity. Much research is being carried out in ABS control systems on various issues and challenges and control methods have been improving over period of time with advancement in technology. Many challenges in ABS are still unresolved and high scope of improvement in the performance of ABS in rough road condition is needed to ensure the safety of drivers and occupants travelling in rough roads. With the high demand of SUVs in today's world and advancement in high-speed cars, safety during panic braking condition is a major concern especially in off-road conditions. On basis of various ABS performance parameters discussed in this paper, an attempt to develop an effective ABS controller for rough road condition will be done considering the latest advancements in soft computing.

References

Adcox, J., Ayalew, B., Rhyne, T., Cron, S., Knauff, M.: Interaction of antilock braking systems with tire torsional dynamics. Tire. Sci. Technol. TSTCA **40**(3), 171–85 (2012)

Aly, A.A., Zeidan, E.S., Hamed, A., Salem, F.: An antilock-braking systems (ABS) control: a technical review. Intell. Control Autom. **2**, 186 (2011)

Bauer, H., Bosch, R.: Driving-Safety Systems. SAE International (1999)

Becker, C.M., Els, P.S.: Profiling of rough terrain. Int. J. Veh. Des. **64**, 240–261 (2014)

Botha, T.R.: High speed autonomous off-road vehicle steering unpublished MEng Thesis. University of Pretoria, Pretoria, South Africa (2011)

Daiss, A., Kiencke, U.: Estimation of vehicle speed fuzzy-estimation in comparison with Kalman-filtering. In: Proceedings of the 4th IEEE Conference on Control Applications, pp. 281–284 (1995)

Day, T.D., Roberts, S.G.: A simulation model for vehicle braking systems fitted with ABS. No. 2002-01-0559. SAE Technical Paper (2002)

Drakunov, S., Ozgiiner, U., Dix, P.: ABS control using optimum search via sliding modes. IEEE Trans. Control Syst. Technol. **3**(1), 79–85 (1995)

Fargione, G., Tringali, D., Risitano, G.: A fuzzy-genetic control system in the ABS for the control of semi-active vehicle suspensions. J. Mechatron. **39**, 89–102 (2016)

Gerotek Test Facilities, Available at: Gillespie, T.: Fundamentals of vehicle dynamics. Warrendale, PA: Society of Automotive Engineers Inc (1992)

Hamersma, H.A., Els, P.S.: Improving the braking performance of a vehicle with ABS and a semi-active suspension system on a rough road. J. Terramech. **56**, 91–101 (2014)

Hamersma, H., Els, P. B., Becker, C., Savitki, D., Heidrich, L., Hopping, K.: The effect of controllable suspension settings on the ABS braking performance of an off-road vehicle on rough terrain. 7th Americas Regional Conference of the ISTVS. Tampa (FL) (2013)

Hamersma, H.A., Schalk Els, P.: Improving the braking performance of a vehicle with ABS and a semi-active suspension system on a rough road. J. Terramech. **56**, 91–101 (2014)

Heidrich, L., Shyrokau, B., Savitski, D., Ivanov, V., Augsburg, K., Wang, D.: Hardware-in-the-loop test rig for integrated vehicle control systems, pp. 683–688. Adv. Automot., Control (2013)

International Organization for Standardization, International Organization for Standardization ISO 8606: Mechanical Vibration–Road Surface Profiles (1995)

Jaiswal, M., Mavros, G., Rahnejat, H., King, P.D.: Influence of tyre transience on antilock braking. Proc. Inst. Mech. Eng. Part K: J. Multi-Body Dyn. **224**, 1–17 (2010)

JATO (2016) "Global Car Market, New car sales 2016-Q1", http://www.jato.com/wcontent/uploads/2016/06/JATO-Global-Car-Sales-2016-Q1.pdf

Jiang, F., and Gao, Z.: An adaptive nonlinear filter approach to the vehicle velocity estimation for ABS. In: Proceedings of the 2000 IEEE International Conference on Control Applications, pp. 490–495 (2000)

Niemz, T., Winner, H.: Reduction of braking distance by control of active dampers. Yokohama, Japan: FISITA World Automotive Congress (2006)

Ozdalyan, B., Blundell, M.V.: Anti-lock braking system simulation and modeling in ADAMS. International conference on Simulation 1998, IEEE Explore (1998)

Penny, W.C.W., Els, P.S.: The test and simulation of ABS on rough, non-deformable terrains. J. Terramech. **67**, 1–10 (2016)
Reul, M., Winner, H.: Enhanced braking performance by integrated ABS and semi-active damping control. In: Proceedings of the 21st (esv) International Technical Conference on the Enhanced Safety of Vehicles. Stuttgart, Germany (2009)
Satoh, M., Shiraishi, S.: Performance of antilock brakes with simplified control technique. SAE Pap. 830484 (1983)
Stan, M., Precup, R.E., Paul, A.S.: Analysis of fuzzy control solutions for antilock braking systems. J. Control Eng. Appl. Inform. **9**(2), 11–22 (2007)
Theron, N., Els, P.: Modelling of a semi-active hydro-pneumatic spring damper unit. Int. J. Veh. Des. **45**(4), 501–521 (2007)
Thoresson, M.J., Els, P.S., Theron, N.J., Uys, P.E.: The ride comfort vs. handling compromise for off-road vehicles. J. Terramech. **44**, 303–317 (2007)
Thoresson, M.J., Uys, P.E., Els, P.S., Snyman, J.A.: Efficient optimisation of a vehicle suspension system, using a gradient-based approximation method. Part 1. Mathematical modeling. Math. Comput. Model. **50**(9), 1421–1436 (2009)
Van der Jagt, P., Pacejka, HB., Savkoor, AR.: Influence of tyre and suspension dynamics on the braking performance of an anti-lock system on uneven roads". In: Proceedings of 2nd International EAEC Conference on New Developments in Powertrain and Chassis Engineering. IMechE C382/047, Strasbourg, France (1989)
Wang, W.-Y., Chien, Y.-H., Li, I.-H.: An On-Line Robust and Adaptive T-S Fuzzy-Neural Controller for More General Unknown Systems. Int. J. Fuzzy Syst. **10**, 33–43 (2008)
Watanabe, M., Noguchi, N.: A New Algorithm for ABS to Compensate for Road Disturbances. SAE Pap. 900205 (1990)

Part II
Safety Analytics

Chapter 5
Data-driven Mapping Between Proactive and Reactive Measures of Occupational Safety Performance

Abhishek Verma, Subit Chatterjee, Sobhan Sarkar and J. Maiti

Abstract This study aims to analyse the incident investigation reports logged after the occurrence of events from an integrated steel plant and map it with proactive safety data. From the narrative text describing the event, this study has attempted to unfold the hazards and safety factors present at the workplace. Text document clustering with expectation maximization algorithm (EM) has been used to group the different events and find key phrases from them. These key phrases are considered as the root causes of the reported events. This study shows how the mapping of the safety factors from both proactive safety data and incident reports can help in the improvement of safety performance as well as better allocation of resources. The study points out specific areas to the management where improvements are needed. The mapping also indicates the areas of improvement made by the constant effort of safety practitioners.

Keywords Incident reports · Proactive safety data · Text document clustering

5.1 Introduction

Steel is essential for the development of any economy because its usage ranges from household products to complex industrial and defence machinery. Major industrial economies are defined and strengthen the growth of strong steel industries. Indian economy is also dependent on the growth of steel industries.

A. Verma (✉) · S. Chatterjee · S. Sarkar · J. Maiti
Department of Industrial & Systems Engineering, IIT Kharagpur, Kharagpur, India
e-mail: abhishekverma.cs@gmail.com

S. Chatterjee
e-mail: subitme16@gmail.com

S. Sarkar
e-mail: sobhan.sarkar@gmail.com

J. Maiti
e-mail: jhareswar.maiti@gmail.com

J. Maiti and P.K. Ray (eds.), *Industrial Safety Management*,
Managing the Asian Century, https://doi.org/10.1007/978-981-10-6328-2_5

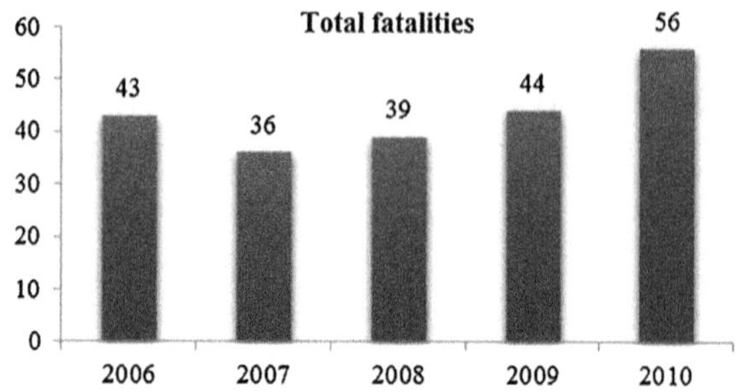

Fig. 5.1 Year-wise distribution of fatalities occurred in major steel producers of India

India holds a key position in world steel production map. Steel sector contributes about 2% to Indian GDP. As per the global scenario, India is the fourth largest producer of crude steel with about 88mt production in the year 2014–2015 (Ministry of Steel, India). Since the production demand is increasing and there is a requirement of building more infrastructure (e.g. workforce, technology, machinery, roads) with the constraints (e.g. space, environmental law, and regulation), which are making the workplace hazardous. The whole process of steel production from raw material to finished products exposes the workers to the wide range of physical (noise, heat, vibration, slip/trip/fall, etc.), chemical (gasses, fuses, etc.) and biological hazards (Jovanovic et al. 2004). In the report of the working group on steel, India for the 12th five-year plan published the year-wise fatalities occurred in major steel producing industries shown in Fig. 5.1. The figure indicates that despite considerable advancement fatalities are increasing over the years. So, ensuring safety at the workplace with keeping the pace of production rate is the topmost priorities for all the industries.

Most of the research efforts in safety management have been focused primarily on analysing and investigating the past accidents but now attention is being directed towards proactive measures to protect the employees and enrich safety culture (Sheehan et al. 2016). Routine safety observation activity can be considered as lead indicators (Dyreborg 2009). In managerial arena there is an increased demand and interest in encouraging firms to use proactive signals rather than relying on lag indicators (incidents/accidents) (Sinelnikov et al. 2015). The importance of proactive safety measures and their relationship to overall business performance are well understood and accepted by senior management. This relationship is pivotal for a proactive strategy in safety management. For this purpose, many of industries are collecting proactive and reactive data.

In the last decade, the data collection in the safety field increases after the introduction of online safety management system database. Root cause extraction from this vast data remains at the core of Safety Management System (SMS) to gather useful information to target the particular system fault because information is captured in both structured and narrative text format. The structured part of data tells about the scenario up to some convincing extent but gets completed by including the narrative text data, at least presumptively. Most of the organizations collect data about the unsafe act, unsafe conditions, and other hazards (lead indicators) in the form of free text. Free text description provides the detail about the hazard, such as the description of the machine, exact location, surrounding condition, to describe the situation prevailing in the plant. Similarly, whenever an incident happens, a brief description of the incident (lag indicator) is reported as incident reports in safety management system (SMS) of the organization in the form of free text. But it also increases the complexity for an analyst to extract the information from that because free text provides the freedom to explain the incident in their words which results in noisy text generation.

To extract hidden information from the vast amount of unstructured data, higher management is searching for methods for taking a smarter decision to improve the safety performance. So, efficient techniques are needed to identify the cause to improve the safety performance in all the safety critical industries. These hidden 'knowledge nuggets' from the large volume of data are practically undetectable using traditional tools and techniques and can be discovered by using advanced techniques of data mining and machine learning (Watson 2008) .

The text mining can support the extraction of lethal factors and pattern from a large volume of data, undiscoverable with the traditional methods. Text clustering analysis is a method for exploration and visualization of textual data, to fully understand the information and the structure of original text documents. Text document clustering is an essential text mining utility which enables the discloser of recurring events. Text clustering analyses have preliminarily shown its performance and usefulness as compared with traditional tools and techniques (Saraçoğlu et al. 2008). For this purpose researchers started using the software with advanced tool and techniques combined with expert analysts (Cleary 2011).

This study introduces safety data analysis as an application area. Steel plant considered for this study captures the hazards and incident related information using an IT enabled safety management system. In this research text document clustering was utilized to study the incident investigation reports and safety observation data to find out root causes and trend, without any prior assumption of their presence. The main target behind the use of clustering techniques is to explore and identify the weak signals from the free text description of incidents and hazards.

Expectation maximization clustering algorithms were used to group loosely related reports and documents, which explains the process of incidents combining the human, technical and organizational factors from incident reports and hazards events from proactive data. It also identifies interconnected reports and discovers the possible recurring irregularities, to determine the hot spots of high risks.

In this study, we propose a methodology for mapping the proactive and reactive safety measure to improve the overall safety performance of the plant. The rest of the paper is organized as follows. Section 5.2 describes data collection preparation. Section 5.3 presents the conceptual research methodology adopted for this study. Results and discussion are given in Sect. 5.4. Conclusions and limitation of the study are provided in Sect. 5.5.

5.2 Data Collection and Preparation

The current study was conducted using incident investigation (reactive) and safety observation data (proactive) of the iron making (IM) division of an integrated steel plant of India. Any employee involved in incident or witness of any hazardous situation at workplace can report the same to the corresponding supervisor of the department or log in the SMS. Whenever worker notices any unsafe act or unsafe condition that has the potential of causing an accident, these observations are recorded into the SMS of the organization. The safety observation data collected under four categories: (i) improper tools and equipment, (ii) rules and procedures, (iii) personal protective equipment (PPE), and (iv) orderliness. The brief description of the hazardous behaviour or condition is noted as free text to narrate the complete scenario. This free text in the form of 'brief description' is an unexplored area and our primary focus for our study. Similarly, whenever incidents happen, employee logs every detail of the incident in the SMS. The incidents are mainly reported under three incident categories: (i) injury/property damage, (ii) medical cases, and (iii) near-miss incidents. These incidents may end up in different impacts. Fifteen different impacts for the various incidents were listed in the incident reports which are: (i) equipment property damage, (ii) derailment, (iii) first aid, (iv) fire, (v) LTI, (vi) exgratia, (vii) toxic Release, (viii) uncontrolled environment, (ix) fatality, (x) medical ailment (major), (xi) medical ailment (minor), (xii) foreign body, (xiii) radio activity, (xiv) death, and (xv) injury on duty. The complete process of incident is narrated in free text form to that can help practitioners in getting insights that are primarily responsible.

Both the safety data for IM division were collected for 17 months (April 2014–August 2015). Figures 5.2 and 5.3 shows the distribution of safety observation and incident data respectively. Figure 5.2 shows that most the observations were made for issues related to orderliness (38%) and tools and equipment (29%).

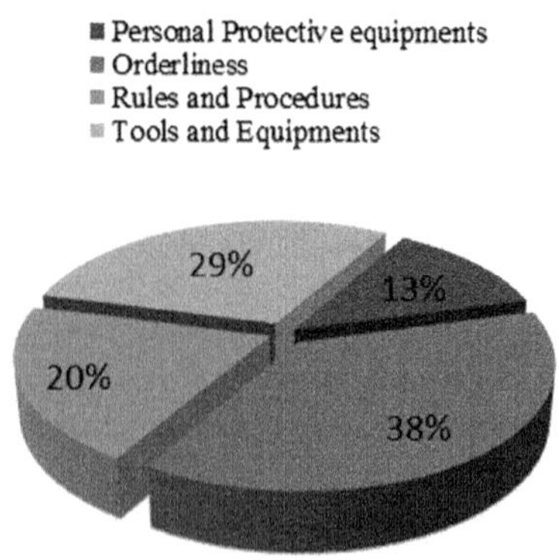

Fig. 5.2 Safety observation data distribution

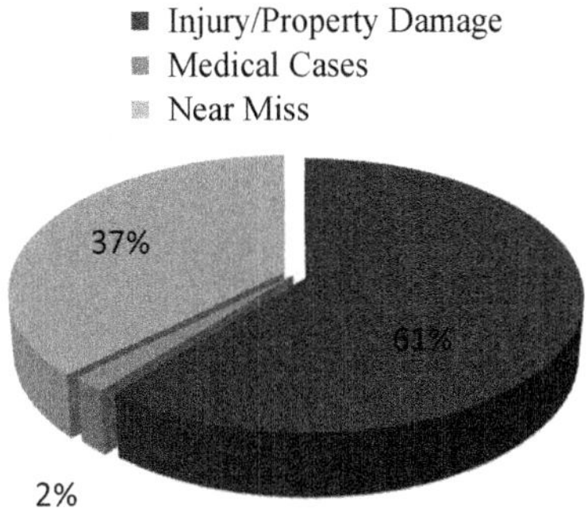

Fig. 5.3 Incident data distribution

Figure 5.3 shows that accidents (61%) are logged more than near-miss (37%) cases. This indicated that worker might not be aware of importance of near-miss reporting.

This study focuses on analysing the descriptive text data. Unfortunately, this section of the report contains lots of inconsistency. So, for the success of any data mining or text mining algorithm, data preparation stage is a deciding factor to extract the quality information (Freitas 2002). Almost 80% time is utilized in pre-processing and preparing the data in any data mining or machine learning project (Zhang et al. 2003). In text mining, data pre-processing is required to remove noise and irrelevant information for improvement in the quality of data (Rajman and Vesely 2004). Data preparation involves various issues related to text data such as a spelling error, non-vocabulary words (Hindi words), incomplete information, irrelevant information (names and address). Duplicate data rows were removed using remove duplicate tool of MS Excel itself. Misspelling and irrelevant shorten text was removed using MS Excel function and manual review.

5.3 Methodology

After data preparation, the variables of interest are extracted from incident reports for further analysis. In our study, the variables of interest are 'incident category' & 'brief description of incident' from incident data and 'observation category' & 'brief description of observation' from safety observation data. In the narrative text data analysis stage, text document clustering technique is utilized to extract the hidden factors in the form of descriptive terms. The descriptive terms from observation data can act like proactive (lead) indicators and descriptive terms form incident data will act as root causes or lag indicators. Then we will try to map all the lagging indicator to corresponding lead indicator. The link will indicate that the lag indicator was already observed as a hazard during safety observation visits. Unlinked lag indicator indicates that there are almost no hazardous observations (lead indicators), logged in past that can be treated as the causal factors.

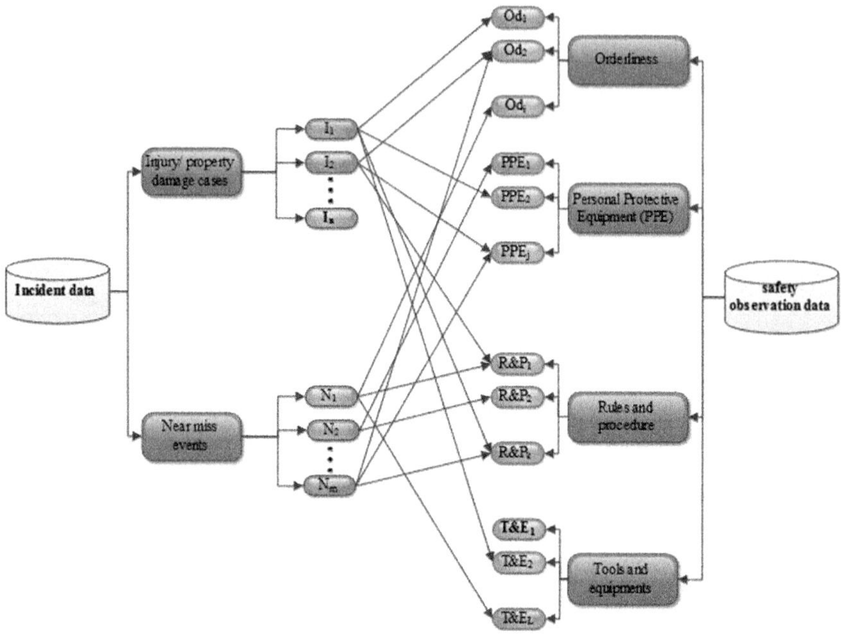

Fig. 5.4 Visualization of mapping the root causes of accident data with safety observation data

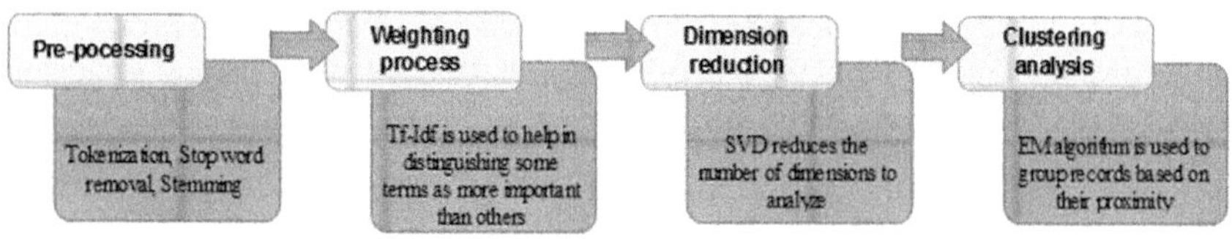

Fig. 5.5 Schematic diagram for text document clustering

Unlinked lead indicator suggests that those hazardous events were successfully mitigated and not converted into any kind of consequences. A visualization of mapping the root causes of incident data with safety observation data has been shown in Fig. 5.4.

The current study utilized SAS text miner software as a text mining tool. The steps performed for text document clustering are shown in Fig. 5.5.

5.4 Results and Discussion

The incident data collected from the SMS had injury/property damage, medical cases and near-miss incidents. Medical cases consist of only 2% of the total incidents. So, medical cases were combined with injury/property damage data for text clustering. Text clustering was performed for accidents (injury/property damage, medical cases) and near-miss incidents separately. With the help of expert advice having wide knowledge of safety domain in steel plant, the root causes/keywords have been inferred from the clustering result. Keywords from the incident clustering result will explore the root causes of accidents. Clustering of injury/medical cases gave 14 clusters and clustering of near-miss data gave eight clusters. Some of the clusters have been combined due to redundancy. Figures 5.6 and 5.7 show the 14 clusters for 'injury/property damage + medical cases' and seven clusters for near-miss incident cases since one cluster of near-miss is redundant. Due to lack of space, the detailed cluster is not provided in the paper (Table 5.1).

It can be seen from the clustering output of injury/property damage + medical cases that wagon derailment cases appear in many cluster (e.g. sinter plant wagon derailment, wagon derailment, quenching loco derailment, and loco torpedo derailment). So these clusters have been clubbed into one broad cluster of 'derailment'. Similarly, 'motorcycle accident' and 'vehicle skidding' clusters are

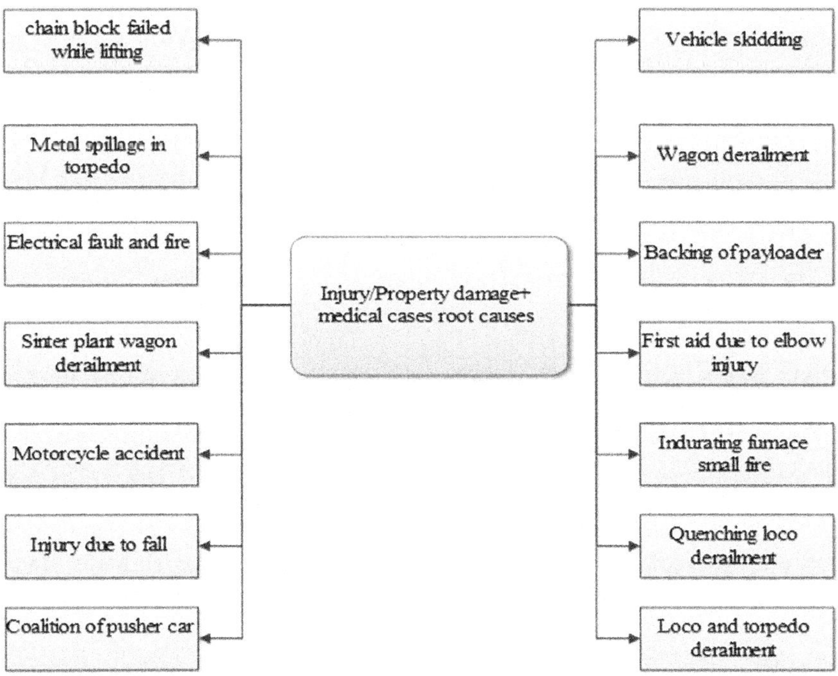

Fig. 5.6 Root causes of injury/property damage + medical cases

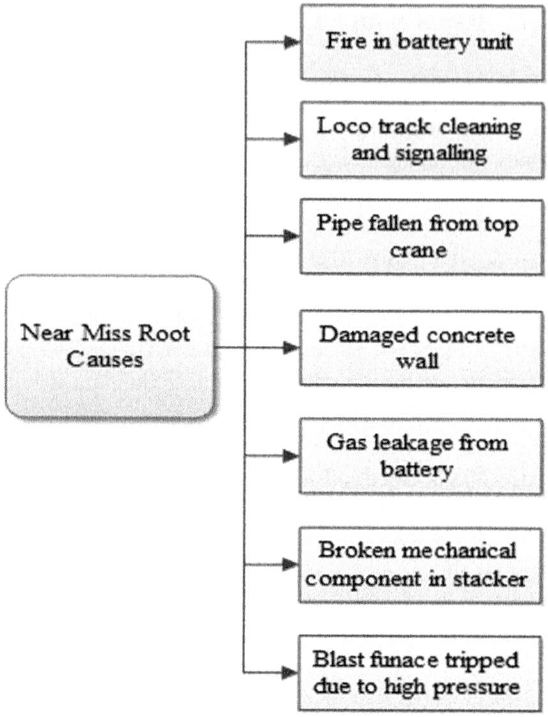

Fig. 5.7 Root causes of near-miss incident

Table 5.1 Number of Text document clusters found for different categories

Sl. No.	Type of data set	Categories	Number of clusters
1	Reactive	Injury/property damage + medical case	14
2		Near-miss	7
3	Proactive	Orderliness	22
4		Personal protective equipment	6
5		Rules and procedure	12
6		Tools and equipment	16

closely related and has been clubbed into one general cluster of 'motorcycle accident'.

Similarly, the safety observation data collected from the SMS had subcategories such as orderliness, personal protective equipment, rules and procedures and tools and equipment. Text clustering was performed on each of these different subcategories separately with the same system settings as used in the incident clustering. Keywords from the clustering result will give us insight into the hazardous situations noted by safety officials during their site visit. However, it was not always possible to decipher keywords from all the clusters. So those clusters have been neglected.

5.4.1 *Mapping of Incident Root Causes Versus Safety Observations*

To identify whether the underlying root causes bringing information about the incidents had been identified as potential risks during safety observation of workplaces, detailed mapping of the root causes of incidents against the lead indicators of hazards noted during safety observation has been done as shown in Fig. 5.8 for injury/property damage + medical cases and in Fig. 5.9 for near-miss incidents.

Few of the underlying root causes such as electrical fault, injury due to fall, coalition of the pusher car in the battery unit, wagon derailment have been looked upon during site visits as potential risks. On the other hand, few of the root causes of accidents such as ‘metal spillage in torpedo’, ‘backing of pay loader’, and ‘chain block problem’ were not anticipated as potential hazards during safety visits. The safety authorities overlooked or neglected to see these as potentially dangerous circumstance. Legitimate moves ought to be made so that these underlying root causes are not disregarded any further and preventive moves ought to be made to moderate any conditions emerging from these root causes. Wagon derailment was found as the most basic cause of accidents. But only a few safety observation data were reflecting the root cause of wagon derailment. Few incidents were reported for the failure of chain pulley or frame failure. However, no safety observations were

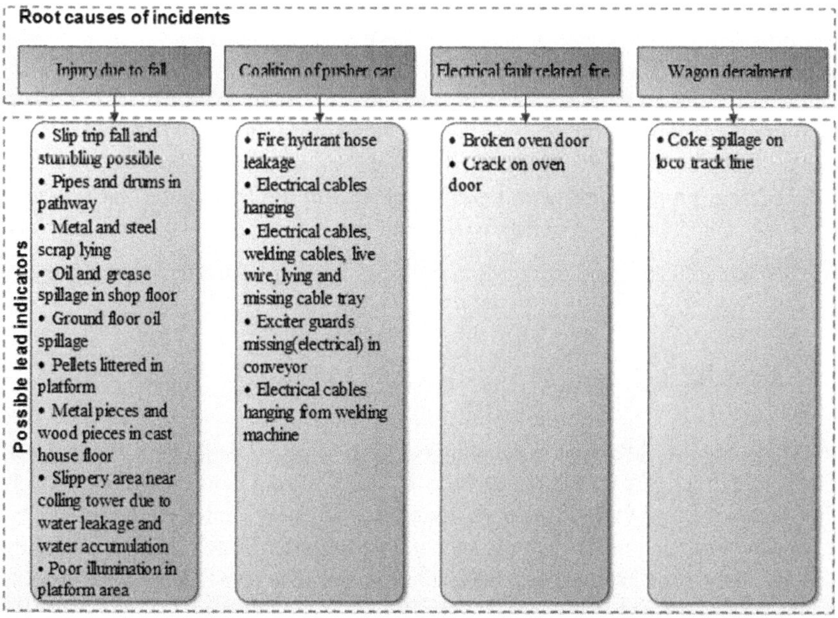

Fig. 5.8 Mapping of root causes of injury/property damage + medical cases with the safety observation data

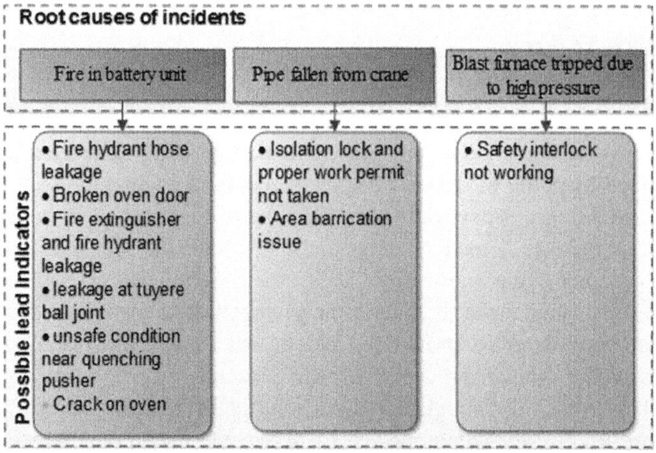

Fig. 5.9 Mapping of root causes of near-miss incidents with safety observation data

logged regarding the checking of chain pulley or other lifting equipment. Third party certification is necessary for lifting equipment. However, none of the safety observations took note of the availability of these certifications of lifting equipment before they were put to work in site.

For the near-miss incidents, many incidents were found due to some issues in the battery unit. The safety observation data show that officials take care of hazardous condition in the battery unit. Leakage of gases due to mechanical failure, broken oven door, the problem with the pusher door was taken care of during the site visit.

Preventive measures were taken to mitigate any fire incident happening in the plant. Fire extinguisher, fire hydrant hose leakages were checked in case of any fire related incidents. Similarly, to avoid falling related incidents, barrication, PPE's of staff working, proper work permit was checked. But the root causes such as loco signalling and damaged concrete wall could not be mapped with and any lead indicators.

5.5 Conclusion

This study tried to consider the mapping between root causes of incidents with lead indicators extracted from the narrative text data. The root causes show the underlying factors behind various incidents and lead indicators shows the focusing area where safety officials are searching for possible hazards. Text document clustering used for analysis of incident and safety observation data provided us with insights into the major accidents happening in the steel plant and their root causes. However, it is not sufficient to showcase all the event description, but it helped in pointing out the areas where the organization needs to improve their safety framework. We have outlined several lead indicators for measurement and monitoring of industrial safety. We are hopeful to conclude here that it can provide practitioners with useful information to support the organization to move away from a focus on lag indicators towards a preventative focus on lead indicators.

Acknowledgements The authors gratefully acknowledge the management of steel plant for their support and cooperation.

References

Cleary, D.: Predictive Analytics in the Public Sector: Using Data Mining to Assist Better Target Selection for Audit. Electron. J. e-Government **9**(2), 132–140 (2011)

Dyreborg, J.: The causal relation between lead and lag indicators. Saf. Sci. **47**(4), 474–475 (2009)

Freitas, A.A.: Data mining and knowledge discovery with evolutionary algorithms, 1st ed. Springer, Heidelberg (2002)

Jovanovic, J., Arandelovic, M., Jovanovic, M.: Multidisciplinary aspects of occupational accidents and injuries. Work. Living Environ. Prot. **2**(4), 325–333 (2004)

Rajman, M., Vesely, M.: From text to knowledge : document processing and visualization : a text mining approach. Text Mining and Its Applications: Results of the NEMIS Launch Conference. Springer Science & Business Media, pp. 7–24 (2004)

Saraçoğlu, R., Tütüncü, K., Allahverdi, N.: A new approach on search for similar documents with multiple categories using fuzzy clustering. Expert Syst. Appl. **34**, 2545–2554 (2008)

Sheehan, C., Donohue, R., Shea, T., Cooper, B., De Cieri, H.: Leading and lagging indicators of occupational health and safety: the moderating role of safety leadership. Accid. Anal. Prev. **92**, 130–138 (2016)

Sinelnikov, S., Inouye, J., Kerper, S.: Using leading indicators to measure occupational health and safety performance. Saf. Sci. **72**, 240–248 (2015)

Watson, R.T.: Data Management, Databases and Organizations. Wiley, New York (2008)

Zhang, S., Zhang, C., Yang, Q.: Data preparation for data mining. Appl. Artif. Intell. **17** March 2015, 375 381 (2003)

Chapter 6
Prediction of Occupational Incidents Using Proactive and Reactive Data: A Data Mining Approach

Sobhan Sarkar, Abhishek Verma and J. Maiti

Abstract Prediction of occupational incidents is an important task for any industry. To do this, reactive data has been used by most of the previous studies in this domain. As an extension of the existing works, the present study has used the underused proactive data coupled with reactive data to establish the predictive models so that the information inherent in both data sets could be better utilized. The main aim of the study is to predict the incident outcomes using mixed data set comprising reactive and proactive data together. Two decision tree classifiers, i.e. classification and regression tree (CART) and C5.0, have been implemented with tenfold cross validation. Furthermore, the ensemble technique, namely adaptive boosting has been implemented to increase the classification accuracy. Results show that boosted C5.0 produces higher accuracy than others for the prediction task. Furthermore, the rules obtained produce the insight of the incidents. The limitation of the present study includes the use of less amount of data and the requirement of experts' domain knowledge for a large span of time. Future scope of the study includes the proper feature selection for preparation of the mixed data set and building the better classification algorithm for better prediction of occurrence of accidents. The present work sets out the potential use of both types of data sources together.

Keywords Occupational incidents · Proactive and reactive data
Decision tree · Ensemble techniques · Rules

S. Sarkar (✉) · A. Verma · J. Maiti
Department of Industrial & Systems Engineering, IIT Kharagpur, Kharagpur, India
e-mail: sobhan.sarkar@gmail.com

A. Verma
e-mail: abhishekverma.cs@gmail.com

J. Maiti
e-mail: jhareswar.maiti@gmail.com

J. Maiti and P.K. Ray (eds.), *Industrial Safety Management*,
Managing the Asian Century, https://doi.org/10.1007/978-981-10-6328-2_6

6.1 Introduction

Despite several layers of safety barriers introduced to enhance occupational safety, accidents of different nature still occur which impart direct and indirect costs to the associated organization (Sanmiquel et al. 2015; Hämäläinen et al. 2006). According to International Labor Organization (ILO), there are more than 317 million accidents reported at work each year, together with occupational diseases leading to more than 2.3 million deaths annually (Rivas et al. 2011) . According to their study more than 48000 workers die yearly in India because of occupational accidents, and almost 37 million cases lead to at least 3 days' absence from work. It was reported in their study that the rate of fatal accidents was 11.4 per 1,00,000 workers and rate of accidents was 8,700 per 1,00,000 workers in India.

Realizing the statistics of consequences of occupational accidents, all over the world, safety in occupation needs to be prioritized and ensured to save human life. A rationale framework in safety management system is therefore required to enhance the accident prevention policy and programs over time (Gautam et al. 2017). However, in most of the cases, it is found that analysis in workplace accident is not rendered a priority (Suarez Sanchez et al. 2011). Attwood et al. (2006) explained the potential contributing factors to this problem as the presence of a fatalistic belief about the accident, i.e. accident will happen. An unfortunate reaction to this situation would be a relaxation of efforts in reduction of accident frequency. In relation with this, ILO makes the following comments: *"Fatalities are not fated; accidents do not just happen; illness is not random; they are caused"* ILO (2003). Therefore, if the working condition is improved, the probability of occurrence of accidents can be decreased. So, the development of a predictive model and its analyses can be helpful which not only predicts the occurrence of accidents in workplaces but also identifies the important factors/contributors behind the accidents so that appropriate preventive measures can be undertaken beforehand.

To analyse the accident scenarios, in most of the cases reactive data are used. Reactive data are usually captured when an event has already been occurred, for example, number of accidents, number of incidents, incidence rate, severity rate. On the other hand, another type of data called proactive data is captured before the occurrence of events, for instances, number of employees working in hazardous conditions, number of workers reporting stress at work, number of training of employees, etc. Most of the previous research works used reactive data to build a model for predicting the occurrence of future incidents. Using historical data, the models are usually built, and trained to predict the future occurrences of accident.

But, the problem in the analysis of using reactive data is that while the historical data is being analysed, the progression of the next accident is already made. This delay in analysis creates a blind spot during which analysts have no insight into what factor is leading to incidents. Proactive data analysis allows the safety management to make changes by observation of developments of factors occurring in real time. Proactive data collected from leading indicators can provide a warning sign before accident occurs, which often remains in hindsight of incident outcomes. Prior to an accident, such signs or "weak signals" are realized only as noise. Therefore, it is important to identify those signals from noise that could help in predicting accidents. There is a common belief that the general "leading indicators" can capture the increasing risk of an accident. While some common indicators may be useful for early identification of level of risk, efforts in a large amount have not been made so far (Khawaji 2012). The reason behind the lack of progress may be the fact that the industry-wide indicators do not exist or may not be effective in particular. Moreover, there exists a belief that major accidents do not happen just because of a unique set of proximal, physical events but from the migration of the organization to a state of heightened risk over a period of time as safe guards and controls get relaxed due to conflicting goals and trade-offs (Rasmussen 1997). If this belief is taken as true one, there should lies a way to identify the migration and intervene prior to a loss. Therefore, identification of leading indicators and analysis of proactive data is necessary.

In support with the leading indicators, as opposed to lagging indicators, some previous studies have mentioned the utility incurred by the use of these indicators. For example, Grabowski et al. (2007) described leading indicators as conditions, events, or means that precede an incident and has a prediction value in regard to an incident. They are just like building blocks of the safety culture across an organization (Hinze et al. 2013). On the contrary, there is a growing number of safety professional questions regarding the predictive value of lagging indicators and their capability of avoiding future accidents. Hence, safety (at workplace) can never be guaranteed by relying only on lagging indicators, rather it also demands continuous focus on lagging indicators of past deficiencies, leading indicators of current technical, organization and human conditions, and leading indicators of technical, organization and human processes that can drive safety forward (Reiman et al. 2012). The value of individual indicator may be of no significance if treated in isolation, but in combination becomes important. However, Mearns (2009) argued that indicators do not necessarily replicate reality, but are an attempt to reflect the truth in the form of data. Realizing the significance of indicators, Reiman et al. (2012) stated that the use of the combination of indicators is useful for an organization and can be compared with other organizations.

Therefore, the findings of their research have motivated us to use the combination of both proactive and reactive data together in such a way so that the future instances could be predicted. However, the scope of the present study is limited only within the steel industry. The wide applicability of the concept of the current study is to be further investigated. In order to predict the occurrence of accidents from the data sets, machine learning (ML) approach has been widely accepted and applied now-a-days. Though initially descriptive and analytical parametric modelling procedures such as descriptive statistical methods (Bevilacqua et al. 2010; Larsson and Field 2002; Biddle and Marsh 2002; Salminen 2005), regression modelling (Roudsari and Ghodsi 2005; Lindell 1997) have been widely used in occupational injury analyses to analyse number, rates, frequency of accidents, ML techniques, recently, have gained much popularity in different domains like engineering, medical science, finance, and they render very useful results (Witten et al. 2016; Han et al. 2011; Friedman et al. 2008) due to their some potential benefits including (i) their capability to deal with large dimensional problems, (ii) their flexibility in reproducing the data generation structure irrespective of complexity, and (iii) their predictive and interpretative potential. So far, studies made on occupational analysis show the use of ML techniques in terms of their predictive power (Matías et al. 2008) and explanatory capacity (Martín et al. 2009). These methods, based on historical data from incident reports, or interview with employees ensure their advantages over conventional statistics in terms of predictive functions and importance of predictors with a bearing on incident outcomes. However, ML techniques, recently popularized, have been used in occupational accident analysis on a limited basis.

In the available literature on the use of ML techniques in analysing occupational accidents, there are basically, two types of approaches used for classification tasks and predictive modelling of incidents: (i) non-tree-based approach like support vector machine (SVM) (Sarkar et al. 2016; Yang et al. 2014; Marucci-Wellman et al. 2017; Chi et al. 2016), k-nearest neighbour (k-NN) (Chen et al. 2010), artificial neural network (ANN) (Bevilacqua et al. 2010; Fragiadakis et al. 2014; Goh and Chua 2013), Bayesian Network (Sarkar et al. 2016) and (ii) tree-based approach i.e., decision tree (DT) (Sanmiquel 2015; Sarkar et al. 2016; Chen and Luo 2016). Among the two approaches mentioned, the later one often shows better performance than the former, though it is subject to testing on the data set being used. The usefulness as well as popularity of DTs is basically due to the fact that they do not require any assumption regarding the distribution of the attribute values; neither do they necessitate any transformation of variables. Moreover, they can process continuous as well as categorical attributes simultaneously. They can be equally useful for classification and regression type of problems (Konda 2010). Pruning, and ensemble, associated with DTs, are powerful methods to enhance its classification accuracy. Due to its higher predictive power, DTs have been applied successfully in diverse fields like medicine (Rahman and Fazle 2011; Oztekin et al. 2011), social sciences (Olson et al. 2012), business management (Aviad and Roy 2011; Chen et al. 2011), construction engineering and management (Shin et al. 2012; Sikder and Munakata 2009), and steel manufacturing (Sarkar et al. 2016a, b, c).

Based on this ground, our present study aims to use both the indicators together (leading and lagging) in terms of proactive and reactive data with an attempt to predict the reactive instances (incident outcomes) from them. Till date, as per authors' knowledge, both proactive and reactive data have not been used in prediction of occurrence of incidents using machine learning (ML) techniques. Our study has addressed this gap and tried to contribute in the existing body of literature. Thus, the main objective of the paper is set to use both proactive and reactive data together to predict the occurrence of accidents in future.

The rest of the paper is organized as follows: in Sect. 6.2, the proposed methodological flowchart is depicted which is followed by a short description of all the methods used in this study. Section 6.3 describes a case study from an integrated steel manufacturing plant. Results and discussions are presented in Sect. 6.4. Finally, in Sect. 6.5, conclusion with future scopes of the study is presented.

6.2 Methodology

In this study, two data sets are collected from the steel industry; one is reactive, i.e. investigation report, and the second one is proactive which is inspection based. After obtaining the data sets, these two are merged based on the link between each observation of inspection and its subsequent logging of incidents in a particular location (here division) at a definite time period. After merging the data sets, a new mixed data set is created which comprises both reactive and proactive instances. All four decision tree algorithms are applied on the data sets and the best model is evaluated in terms of prediction power. The total process is depicted in Fig. 6.1.

6.3 Case Study

6.3.1 Data Set Generation

There are two data sets, one proactive and another reactive, were collected from the electronic database of an integrated steel making organization during the year 2010 to 2013. The reactive data set, which is called inspection data, represents the

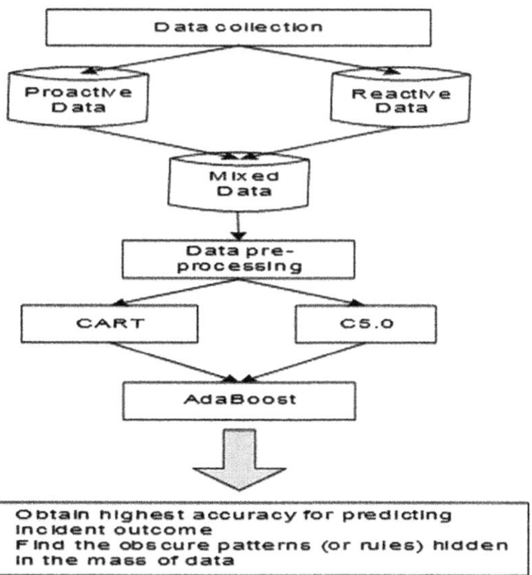

Fig. 6.1 Flowchart of the proposed methodology

observation by safety personnel during his visit into the working area. If he finds anything unusual or deviated from the standard operating procedure (SOP), he logs the observation indicating a certain level of potentiality of hazards. On the contrary, in other data set, which is considered as the reactive one, the observation is recorded after the incident is taken place as investigation output. After obtaining the two sets of data, a mixed data set consisting of 7526 data points is created from them maintaining the link between the observatory details in proactive data set and investigated details in reactive data set for a certain division at a certain period of time. The attributes consider from the proactive data are *month, department, section, observation related to, type of activity, observation category and observation type*; whereas the attributes taken for the reactive data are *primary cause, working condition, incident type and incident outcome.*

6.3.2 Data Preprocessing

Once the data set is generated, it is cleaned and preprocessed with the help of R, an open-source software. This step of data preprocessing includes removal of missing values and duplicate entries from the data sets. It also excludes the data points from the reactive data set having inconsistencies with proactive data description. Once the preprocessing is completed on the mixed data set, a descriptive analysis of the attributes was carried out to present the frequency distribution of incidents for (injury, and non-injury) of each attribute. Importance of each of the features or predictors towards the response attribute is calculated by feature selection algorithm (here, Boruta algorithm). The reason behind selecting this algorithm is that it can capture all features which are in some circumstances weakly or strongly relevant to the outcome variable, whereas other traditional feature selection algorithms follow minimal optimal method where they rely on small subset of features yielding

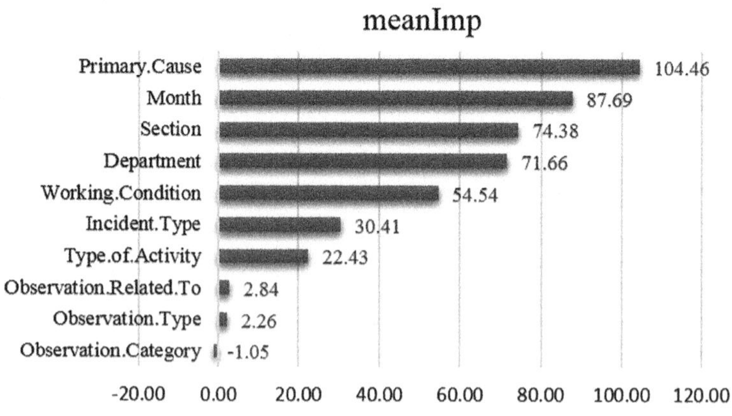

Fig. 6.2 Mean importance of predictors obtained from Boruta algorithm

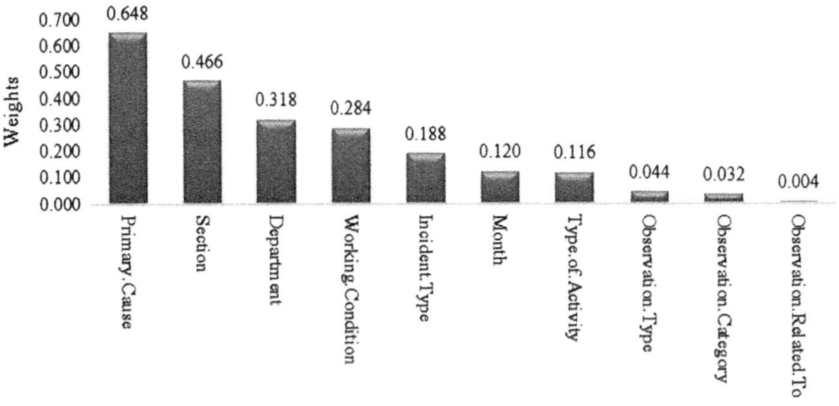

Fig. 6.3 Feature importance weights from Chi-square algorithm

minimal errors on a chosen classifier. Furthermore, Chi-square feature selection algorithm (see Fig. 6.3) is also applied to validate the result obtained from the Boruta algorithm (see Fig. 6.2). The same set of features is found to be important towards outcome of the model. After that, machine learning techniques have been used to the data sets to predict the incident outcome.

6.3.3 Methods Used

In literature, there are many DT inductive learning algorithms available, such as Iterative Dichotomiser (ID3) (Quinlan 1983,1986), C4.5 (Quinlan 1993), C5.0 (Ville 2006), Classification and Regression Trees (CART) (Breiman et al. 1984) and Chi-squared Automatic Interaction Detection (CHAID) (Kass 1980), Quick, Unbiased, Efficient Statistical Tree (QUEST) (Bevilacqua et al. 2010). The performance of these algorithms mainly depends on the two criteria; (i) splitting criteria to create parent and child nodes, and (ii) the number of splits in the child nodes. Therefore, in this study, it is intended to analyse our data set to predict the incident outcome with two different decision tree algorithms-CART, and C5.0. Thereafter, in order to increase the accuracy of prediction of decision tree classifiers, adaptive boosting (i.e. AdaBoost) has been used.

6.3.3.1 C5.0 Algorithm

This algorithm developed by Quinlan is based on entropy indicating the amount of information that a signal contains. This DT algorithm usually selects the attribute having highest amount of information inherent within the data set as a root node. Then, based on the splitting criteria of the root node, the data is separated into child nodes. The stopping criterion of splitting is the situation when most of the data points within a branch belong to a same class, and consequently, the node is labelled with that class. This process continues recursively for all branches of the decision tree until all cases in each of the branches lead to a creation of a leaf node or certain stopping conditions are met. This algorithm facilitates boosting and cross-validation operation (Ville 2006). The main advantage of this algorithm is that it is faster than C4.5; has better effectiveness in memory; can build smaller trees; can provide boosting operation for obtaining improved accuracy; can provide weights to different variables; and can efficiently perform noise reduction, i.e. winnowing. This method uses "information gain ratio" to estimate split at each internal node of the tree. The metric information gain computes the reduction in entropy by a split. Using this information gain value, the test is chosen based on subdivision of the data ensuring the maximization of decrease in entropy of the descendant nodes.

6.3.3.2 CART Algorithm

Classification and regression tree (CART) analysis is a recursive approach that can be used for both regression and classification task by the generation of trees (Breiman et al. 1984). First, it splits the whole feature space into non-overlapping regions and then predicting the most likely value of the dependent variable within each of the regions. Basically, based on the type of response/dependent variable, a classification tree is built by utilizing impurity to split the Gini Index of diversity. One problem that arises here is the increase of complexity of the functionality of the algorithm. Thus, in order to reduce the complexity, using cost-complexity parameter, tree is pruned which takes care both the criteria; (i) precision and (ii) complexity in terms of processing speed and number of nodes. For the complete illustration of the method, interested authors are requested to refer Breiman et al. (1984).

6.3.3.3 Adaptive Boosting

Boosting is an ensemble learning method. Opposed to bagging, another ensemble technique, it is designed to boost a "weak" learner into a "strong or powerful" one (Hou et al. 2015; Theodoridis and Koutroumbas 2010) so that higher degree of accuracy can be achieved. Functionally, it uses a series of base classifiers with different weights by taking different subsets of the training set. The weighting distribution at each iteration identifies the misclassified cases by the previous base classifier. Finally, boosting classifier is generated by averaging the weight values provided to each of the base classifier. Adaptive boosting (i.e. AdaBoost) is a popular algorithm from the family of boosting algorithms. The training process of it can be found in some previous works (Freund and Schapire 1997; López and Abellán 2013).

6.4 Research Results and Discussion

In this section, the results from the feature selection and decision tree classifiers are shown and discussed. First of all, the features or attributes used in this study for the prediction of incident outcomes are analysed and their importance is checked towards prediction of the outcome variable. Two techniques, namely Boruta feature selection and chi-square are employed to check predictors' importance. The results reveal that attributes like 'Primary Cause', 'Month', 'Section', 'Department', 'Working condition', 'Incident type' and 'Type of activity' have higher prediction value. Among them, 'Primary Cause' is found to be the most powerful/strong predictors towards the outcome which is supported by the results of both algorithms. The rest of the attributes like 'Observation related to', 'Observation type', and 'Observation category' are found to be the weak predictors of the outcome variable. This may be due to the randomness of the data point for the three attributes that does not help in the prediction. For the less number of predictors used in this study, all of them are kept for analysis of decision tree prediction.

There are two base decision tree algorithms, i.e. CART and C5.0, have been applied on the mixed data sets to predict the incident outcome as response variable. Furthermore, to check the utility of the data base in the prediction by DTs, proactive and reactive data are analysed separately. In order to increase the accuracy of classifiers, Adaptive Boosting (i.e. AdaBoost) has been implemented on each of the base DTs.

The decision tree generated by using the C5.0 algorithm is found to be the better classifier than CART in terms of prediction accuracy. Using tenfold cross validation, C5.0 produces 98.71% accuracy in normal condition (i.e. without boosting), whereas CART produces 91.1% accuracy while mixed data set is used. It is also observed from the Fig. 6.4 that C5.0 algorithm predicts more accurately using mixed data than either of other data sets. In addition, CART also shows the same result by showing the superior classification accuracy obtained while mixed data is used. The results reveal that instead of using only reactive data sets for prediction or classification tasks if the mixed data sets could be used, prediction capability of the classifiers can be increased. In order to increase the accuracy, AdaBoost on DT algorithms has been used for all the three types of data sets using tenfold cross validation. Results reveal that C5.0 algorithm outperforms CART marginally in terms of accuracy (98.88% for boosted C5.0 and 98.71% for boosted CART). While checking the accuracy of the classifiers using other two types of data sets, it is observed that C5.0 produces higher accuracy in prediction using mixed data set than other type of data sets. Therefore, it can be concluded that boosted C5.0 predicts higher than CART using mixed data. In addition, prediction using mixed data set is more accurate than using either of other data sets.

DT algorithm is important for interpreting the results of prediction of target variable in terms of a logical set of rules. Once DT is constructed, rules can be easily generated. However, while ensemble technique is used on single base DT classifier, it is very hard to interpret the rules from ensembled trees or a collection of trees.

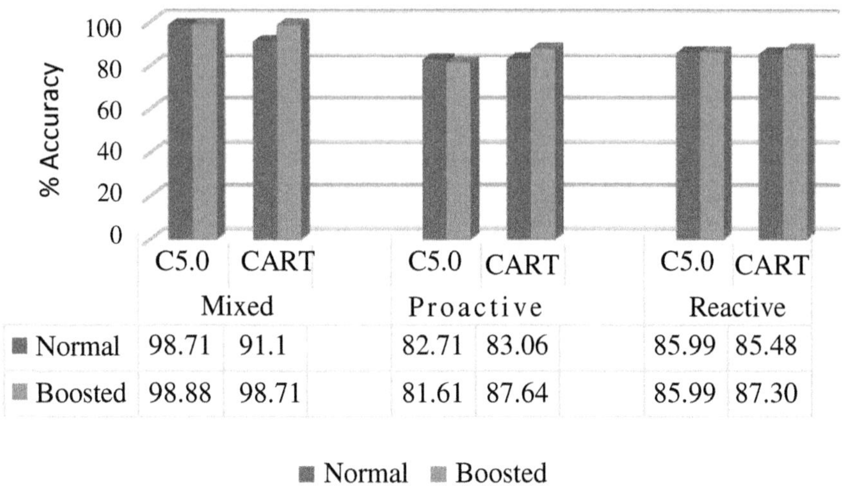

Fig. 6.4 Accuracies of two DT algorithms used in three types of data sets

Thus, for understanding purpose, rules generated from C5.0 algorithm are discussed only in this paper. Each of the rules indicates the injury or non-injury cases as consequences.

There is a set of such 108 rules generated from C5.0 algorithm of which only top 24 rules explaining most frequent injuries are tabulated in Table 6.1. Among them, one of the important rules occurring most frequently is skidding which occurs in blast furnace and sinter plant area. This may be due to the fact that pallets, which are spherical in shape and are used in those area, are often found lying down on floor area which causes skidding. Another rule explains that during lifting tools and tackles, workers get injured mostly. This may be due to the improper or no training to the workers while in operating with lifting tools such as electric chain hoists, eye bolts, D-shackles, chains in the workplace. Lack of carefulness during works may also result in injury. Similarly, another interesting rule is during the month of April and December, injuries occur most frequently in Blast furnace area. If the management take care of the factors, which are identified as decision rules, leading to accidents in workplace, it becomes easier for the organization to reduce the number of accidents. Thus, the C5.0 algorithm has more potential to explore the hidden rules from the accident data, and help the organization in decision making in risk management.

However, the present study has also some of the limitations. One of them is data preprocessing technique. In this stage, some observations, though realized as important ones, are deleted from the data set due to the existence of missing values in some of the attributes which, in turn, leads to suppressing some rules. Another important limitation of the study is that rules cannot be interpreted while ensemble of DTs is applied on single base DT classifier. In this present study, two DT algorithms, i.e. C5.0 and CART, have been used. After that, Adaptive Boosting, applied on both of the data sets, shows that boosted C5.0 algorithm outperforms boosted CART. But, the rules cannot be interpreted due to the huge number of rules from an assembly of decision trees.

6.5 Conclusion

In the present study, a unique approach has been adopted to predict the reactive instance using mixed data, i.e. both proactive and reactive data. Initially, the mixed data set has been prepared using from both inspection report and investigation logged incidents report. Once the mixed data set is prepared, machine learning algorithm is applied to predict the reactive instance, i.e., incident outcome. Decision tree algorithms, i.e. C5.0 and CART, have been used. Adaptive Boosting has been applied in order to increase the level of accuracy of the classifiers. Boosted C5.0 algorithm is found to have higher accuracy (98.88%) than others. To avoid the model over fitting, tenfold cross validation has been performed. In addition, the results also explore the fact that prediction using mixed data is useful than that of single data set like either proactive or reactive data. The rules generated from C5.0 algorithm produce another insight towards accident occurrences that additionally help decision makers take judicious decisions to reduce the accidents at workplaces. Some primary causes like skidding, lifting tools and tackles, STF, process incidents are frequently causing accidents, thus injuries, in the workplaces like sinter plant, blast furnace areas. However, the concept used for prediction of reactive instances using mixed data type is new altogether which requires more research to validate it. Therefore, future scopes of the present study can encompass the creation of mixed data to predict future instances. The more the relevant predictors are available, the more will be the prediction strong. Another important study that can help to predict occurrences of accidents from proactive data only can also be carried out that ultimately helps decision makers avoid the dependence completely on reactive data. Preprocessing operations can also be done carefully to enhance data quality using advanced algorithms which can eventually ensure higher predictive capability of the model.

Acknowledgements The authors gratefully acknowledge the management of steel plant for their support and cooperation.

Appendix

(Table 6.1)

Table 6.1 Top 24 rules explaining most frequent injuries

Rules	Injury (%)	No. of cases	% of cases
Primary Cause in ("Skidding") and Department in ("H Blast Furnace" "Sinter Plant")	100	131	1.741
Primary Cause in ("Lifting Tools and Tackles")	100	115	1.528
Primary Cause in ("Slip/Trip/Fall") and Department in ("A-F Blast Furnace") and Month in ("April" "December")	100	105	1.395
Primary Cause in ("Process Incidents") and Working Condition = Group Working and Month in ("September") and Department in ("A-F Blast Furnace" "Coke Plant")	100	87	1.156
Primary Cause in ("Road Incident") and Department in ("H Blast Furnace") and Working Condition = Single Working	100	80	1.063
Primary Cause in ("Slip/Trip/Fall") and Department in ("Sinter Plant") and Month in ("September")	100	74	0.983
Primary Cause in ("Dashing/Collision") and Department in ("A-F Blast Furnace") and Month in ("November")	100	66	0.88
Primary Cause in ("Material Handling") and Department in ("Sinter Plant")	100	54	0.718
Primary Cause in ("Slip/Trip/Fall") and Department in ("Raw Material Handling Group") and Month in ("December")	100	47	0.625
Primary Cause in ("Crane Dashing") and Month in ("June")	100	46	0.61
Primary Cause in ("Road Incident") and Department in ("Raw Material Handling Group") and Month in ("November")	100	43	0.571
Primary Cause in ("Process Incidents") and Working Condition = Group Working and Month in ("July")	100	42	0.558
Primary Cause in ("Structural Integrity") and Month in ("June")	100	39	0.518
Primary Cause in ("Material Handling") and Department in ("Coke Plant") and Month in ("September")	100	34	0.452
Primary Cause in ("Dashing/Collision") and Department in ("G Blast Furnace") and Month in ("November" "October")	100	33	0.44
Primary Cause in ("Energy Isolation") and Department in ("Coke Plant")	100	33	0.44
Primary Cause in ("Medical Ailment") and Working Condition = Not Applicable	100	25	0.332
Primary Cause in ("Road Incident") and Department in ("A-F Blast Furnace") and Working Condition = Single Working	100	21	0.279
Primary Cause in ("Dashing/Collision") and Department in ("H Blast Furnace")	100	20	0.27
Primary Cause in ("Process Incidents") and Working Condition = Group Working and Month in ("November") and Department in ("Hot Metal Logistics")	100	20	0.266
Primary Cause in ("Gas Leakage") and Month in ("July")	100	15	0.2
Primary Cause in ("Hydraulic/Pneumatic") and Working Condition = Group Working	100	15	0.199
Primary Cause in ("Slip/Trip/Fall") and Department in ("G Blast Furnace" "Hot Metal Logistics")	100	13	0.173

References

Attwood, D., Khan, F., Veitch, B.: Can we predict occupational accident frequency? Process Saf. Environ. Prot. **84**(3), 208–221 (2006)

Aviad, B., Roy, G.: Classification by clustering decision tree-like classifier based on adjusted clusters. Expert Syst. Appl. **38**(7), 8220–8228 (2011)

Bevilacqua, M., Ciarapica, F., Giacchetta, G.: Data mining for occupational injury risk: a case study. Int. J. Reliab. Qual. Saf. Eng. **17**(4), 351–380 (2010)

Biddle, E.A., Marsh, S.M.: Comparison of two fatal occupational injury surveillance systems in the United States. J. Safety Res. **33**(3), 337–354 (2002)

Breiman, L., Friedman, J., Stone, C.J., Olshen, R.A.: Classification Algorithms and Regression Trees. CRC press (1984)

Chen, H., Luo, X.: Severity prediction models of falling risk for workers at height. Procedia Eng. **164**(June), 439–445 (2016)

Chen, W.T., Chang, P.Y., Chou, K., Mortis, L.E.: Developing a CBR-based adjudication system for fatal construction industry occupational accidents. Part I: Building the system framework. Expert Syst. Appl. **37**(7), 4867–4880 (2010)

Chen, J.-H., Su, M.-C., Chen, C.-Y., Hsu, F.-H., Wu, C.-C.: Application of neural networks for detecting erroneous tax reports from construction companies. Autom. Constr. **20**(7), 935–939 (2011)

Chi, N., Lin, K., El-gohary, N., Hsieh, S.: Evaluating the strength of text classification categories for supporting construction field inspection. Autom. Constr. **64**, 78–88 (2016)

De Oña, J., López, G., Abellán, J.: Extracting decision rules from police accident reports through decision trees. Accid. Anal. Prev. **50**, 1151–1160 (2013)

De Ville, B.: Decision Tree for Business Intelligence and Data Mining. SAS Publishing, Cary (2006)

Fragiadakis, N.G., Tsoukalas, V.D., Papazoglou, V.J.: An adaptive neuro-fuzzy inference system (anfis) model for assessing occupational risk in the shipbuilding industry. Saf. Sci. **63**, 226–235 (2014)

Freund, Y., Schapire, R.: A desicion theoretic generalization of on line learning and an application to boosting. Eur. Conf. Comput. Learn. Theory **55**, 119–139 (1997)

Friedman, J., Hastie, T., Tibshirani, R.: The elements of statistical learning. Springer, California (2008)

Gautam, S., Maiti, J., Syamsundar, A., Sarkar, S.: Segmented point process models for work system safety analysis. Saf. Sci. **95**, 15–27 (2017)

Goh, Y.M., Chua, D.: Neural network analysis of construction safety management systems: a case study in Singapore. Constr. Manag. Econ. **31**(5), 460–470 (2013)

Grabowski, M., Ayyalasomayajula, P., Merrick, J., Harrald, J.R., Roberts, K.: Leading indicators of safety in virtual organizations. Saf. Sci. **45**(10), 1013–1043 (2007)

Hämäläinen, P., Takala, J., Saarela, K.L.: Global estimates of occupational accidents. Saf. Sci. **44**(2), 137–156 (2006)

Han, J., Pei, J., Kamber, M.: Data Mining: Concepts and Techniques. Elsevier (2011)

Hinze, J., Thurman, S., Wehle, A.: Leading indicators of construction safety performance. Saf. Sci. **51**(1), 23–28 (2013)

Hou, Y., Edara, P., Sun, C.: Situation assessment and decision making for lane change assistance using ensemble learning methods. Expert Syst. Appl. **42**(8), 3875–3882 (2015)

ILO, "Safety in numbers - Pointers for global safety culture at work" (2003)

Kass, G.V.: An exploratory technique for investigating large quantities of categorical data. Appl. Stat. **29**(2), 119–127 (1980)

Khawaji, I.A.: Developing system-based leading indicators for proactive risk management in the chemical processing industry. Massachusetts Institute of Technology (2012)

Konda, R.: Predicting Machining Rate in Non-Traditional Machining Using Decision Tree Inductive Learning. Nova Southeastern University (2010)

Larsson, T.J., Field, B.: The distribution of occupational injury risks in the state of Victoria. Saf. Sci. **40**(5), 419–437 (2002)

Lindell, M.K.: Occupational safety and health inspection scores predict rates of workers' lost-time injuries. Accid. Anal. Prev. **29**(5), 563–571 (1997)

Martín, J.E., Rivas, T., Matías, J., Taboada, J., Argüelles, A.: A Bayesian network analysis of workplace accidents caused by falls from a height. Saf. Sci. **47**(2), 206–214 (2009)

Marucci-Wellman, H.R., Corns, H.L., Lehto, M.R.: Classifying injury narratives of large administrative databases for surveillance—a practical approach combining machine learning ensembles and human review. Accid. Anal. Prev. **98**, 359–371 (2017)

Matías, J.M., Rivas, T., Martín, J.E., Taboada, J.: A machine learning methodology for the analysis of workplace accidents. Int. J. Comput. Math. **85**(3–4), 559–578 (2008)

Mearns, K.: From reactive to proactive—can LPIs deliver? Saf. Sci. **47**(4), 491–492 (2009)

Olson, D.L., Delen, D., Meng, Y.: Comparative analysis of data mining methods for bankruptcy prediction. Decis. Support Syst. **52**(2), 464–473 (2012)

Oztekin, A., Kong, Z.J., Delen, D.: Development of a structural equation modeling-based decision tree methodology for the analysis of lung transplantations. Decis. Support Syst. **51**(1), 155–166 (2011)

Quinlan, J.: Learning efficient classification procedures and their application to chess end games. Mach. Learn. An Artif. Intell. Approach **1**, 463–482 (1983)

Quinlan, J.R.: Induction of decision trees. Mach. Learn. **1**(1), 81–106 (1986)

Quinlan, J.R., C4. 5: Programs for Machine Learning. Elsevier, Burlington (1993)

Rahman, R.M., Fazle, F.R.: Using and comparing different decision tree classification techniques for mining ICDDR,B Hospital Surveillance data. Expert Syst. Appl. **38**(9), 11421–11436 (2011)

Rasmussen, J.: Risk management in a dynamic society: a modelling problem. Saf. Sci. **27**(2), 183–213 (1997)

Reiman, T., Pietikäinen, E.: Leading indicators of system safety -Monitoring and driving the organizational safety potential. Saf. Sci. **50**(10), 1993–2000 (2012)

Rivas, T., Paz, M., Martín, J.E., Matías, J.M., García, J.F., Taboada, J.: Explaining and predicting workplace accidents using data-mining techniques. Reliab. Eng. Syst. Saf. **96**(7), 739–747 (2011)

Roudsari, B.S., Ghodsi, M.: Occupational injuries in Tehran. Injury **36**(1), 33–39 (2005)

Salminen, S.: Relationships between injuries at work and during leisure time. Accid. Anal. Prev. **37**(2), 373–376 (2005)

Sanmiquel, L., Rossell, J.M., Vintro, C.: Study of Spanish mining accidents using data mining techniques. Saf. Sci. **75**, 49–55 (2015)

Sarkar, S., Patel, A., Madaan, S., Maiti, J.: Prediction of occupational acciden using decision tree approach. In: India Conference (INDICON), 2016 IEEE Annual, pp. 1–6 (2016a)

Sarkar, S., Vinay, S., Maiti, J.: Text mining based safety risk assessment and prediction of occupational accidents in a steel plant. In: 2016 International Conference on Computational Techniques in Information and Communication Technologies (ICCTICT), pp. 439–444 (2016b)

Sarkar, S., Vinay, S., Pateshwari, V., Maiti, J.: Study of optimized SVM for incident prediction of a steel plant in India. In: India Conference (INDICON), 2016 IEEE Annual, pp. 1–6 (2016c)

Shin, Y., Kim, K., Cho, H., Kang, K.-I.: A formwork method selection model based on boosted decision trees in tall building construction. Autom. Constr. **23**, 47–54 (2012)

Sikder, I.U., Munakata, T.: Application of rough set and decision tree for characterization of premonitory factors of low seismic activity. Expert Syst. Appl. **36**(1), 102–110 (2009)

Suárez Sánchez, A., Riesgo Fernández, P., Sánchez Lasheras, F., De Cos Juez, F.J., García Nieto, P.J.: Prediction of work-related accidents according to working conditions using support vector machines. Appl. Math. Comput. **218**(7), 3539–3552 (2011)

Theodoridis, S., Koutroumbas, K.: Introduction to Pattern Recognition: A Matlab Approach. Academic Press (2010)

Witten, I.H, Frank, E., Hall, M.A., Pal, C.J.: Data Mining: Practical Machine Learning Tools and Techniques. Morgan Kaufmann, Burlington (2016)
Yang, K., Aria, S., Ahn, C.R., Stentz, T.L.: Automated detection of near-miss fall incidents in iron workers using inertial measurement units. Constr. Res. Congr. **2014**, 935–944 (2014)

Chapter 7
Determinants of Risk Indices in Hard Rock Mine Using Loglinear Model

Falguni Sarkar, P.S. Paul and Aveek Mangal

Abstract Occupational hazards are undeniably a foremost determinant of inadequate initiative and contribute to collective inequalities in injuries, health, disabilities, and hasty transience. This study aimed to evaluate the roles of occupational hazards and their contributions to the occurrences of injuries among the Indian hard rock mine workers. Injury rates are calculated based on the normalized injury rate (NIR), and a loglinear model was investigated for the identification of causative factors involved mine accident through the cross-classification table. The application of loglinear model expedited the quantitative analyses of mine safety problems in a multivariate state of affairs on a hard rock mine in the eastern part of India. This model illustrates that the variations observed in cell frequencies are not indiscriminate and occupation is a major contributing factor followed by workplace location. Estimation of risk is done by the parameter calculated in this model, and important information about the reduction of the accidents in the underground metal mine is presented.

Keywords Occupational injuries · Accident analysis · Hierarchical loglinear models · Odds ratio · Association

7.1 Introduction

The mining industry has a high incidence of injury among all industry divisions; particularly, underground hard rock mining has been and remains one of the most dangerous occupations in the globe (Kunar and Kumar 2011). In recent years, the

F. Sarkar
Internal Safety Organisation (ISO) and Superintendent of Mines, Narwapahar, Jharkhand, India

F. Sarkar (✉)
Uranium Corporation of India Limited, Jaduguda, Jharkhand, India
e-mail: falguniind31@gmail.com

P.S. Paul · A. Mangal
Department of Mining Engineering, Indian Institute of Technology (Indian School of Mines), Dhanbad 826004, Jharkhand, India

J. Maiti and P.K. Ray (eds.), *Industrial Safety Management*, Managing the Asian Century, https://doi.org/10.1007/978-981-10-6328-2_7

occupational injury rate in underground hard rock mining has been six times higher than that in all other industries (Pal and Khanda 2008). Several studies have been attempted to identify the personal and environmental factors which are predisposed some individuals, rather than the others factors, to be injured at workplace (Maiti and Bhattacherjee 1999). The predisposing factors can be divided into five main groups such as age, experience, occupation, location, and degree of injury, and statistical analysis was performed using statistical packages for Social Sciences package program. In this study, occupational accidents occurred in underground metal mines of eastern part of India between the years 2002–2013 were examined.

7.2 Methods

Accident studies are used to identify common factors contributing to occupational accidents and to give recommendations for accident prevention. The studies conducted on the occurrence of injures in underground hard rock mines have identified a number of variables affecting mine accidents. Based on the published literature (Ghosh and Bhattacherjee 2009; Rahman et al. 2014; Onder et al. 2014) and the accident records, the variables chosen in this study were divided into the five main groups such as occupation, miners' age, experience, location, and degree of injury.

7.2.1 Variables

Based on the analyses by Rahman et al. (2014), Onder et al. (2014), Ghosh (2010), Onder and Adiguzel (2010), the variables chosen in this study are the following: miners' age, experience, occupation, workplace, location, and degree of injury. The following is a brief description of the findings for each variable as it relates to accident and injury analyses.

7.2.1.1 Age

Many studies were conducted to observe the effect of age on accident rate. The NRC (1982) found strong negative correlation between age and disabling injury rate in mining industry. It was concluded that younger miners experience a much higher disabling injury rate than the older miners. Maiti et al. (2001) showed that the injury frequency rate decreases with age, whereas Bhattacherjee et al. (2007) observed the opposite trend. Ghosh and Bhattacherjee (2009) found conflicting results of the younger miners showed no difference in the probability of occurrence of an injury than middle-aged miners. According to Ghosh (2010), younger workers have a higher injury rate because of less job experience, whereas the older workers have high injuries due to age factor. Several studies have reported both positive and negative relationships between age and other factors related to work injury.

7.2.1.2 Experience

Experience has long been debated for their causal influences to work injury, and it represents the volume of time an employee has been engaged in their work. Conflicting results have been reported in the literature regarding this point. Prior studies have shown negative, positive, or no relationship between experience and work injury (Maiti and Bhattacherjee 1999; Onder and Adiguzel 2010). Young age is associated with a lack of experience which predisposes to the occurrence of injuries. It could also be associated with a lack of job, knowledge, and know-how (Ghosh and Bhattacherjee 2009).

7.2.1.3 Occupation

The most hazardous job classifications of researchers' study were roof bolters, loading machine operators, and working supervisors. They classified the miners in four job classifications, namely mobile equipment operators, supervisory personnel, maintenance personnel, and other job personnel. Bhattacherjee and Kunar (2011) found that the supervisors and the maintenance workers were less likely to experience serious injuries than the other job personnel, whereas the mobile equipment operators and other job personnel had the same probability of experiencing serious injuries. So, occupation is one of the important factors for occupational injuries for miners to be considered for analysis.

7.2.1.4 Location

Bhattacherjee and Kunar (2011) examined the relationship between accident, location, and severity of injury, and further, Ghosh and Bhattacherjee (2007) point out that face workers had higher risk of injuries compared to the other workers. This was also expected as they were exposed to more hazardous work conditions, and miners should always work in changing work environment under freshly exposed roof where ventilation and other environmental conditions could be inadequate, resulting in heat, humidity, and slippery floor (Ghosh et al. 2004).

7.2.1.5 Degree of Injury

This variable is important as it measures the injury severity, which is also an indirect measure of the cost of an accident. According to Onder and Adiguzel (2010), the degree of injury of face winner job group had high exposure to accidents in face areas due to falls of ground, whereas mechanic–electrician had high risk due to machinery. However, Onder et al. (2014) conducted a study of degree of injury in occupational injuries which were evaluated with respect to occupation, area, reason, accident time and part of body affected, and the loss of work days due to accidents which is noteworthy (Rahman et al. 2014).

7.2.2 Loglinear Model

Regarding statistical methods, authors suggested loglinear models for analysing cross-tabulated data due to the multifactorial nature of the causes of occupational injuries (Maiti et al. 2001). In this study, loglinear models are of primary use when the variables investigated by the model are all treated as response variables, and therefore, the model demonstrates association between variables of the loglinear model, which is investigated as a statistical representation of the contingency table to quantify the relative importance of various factors and to detect occasions where the effects of these factors can be presumed real or they are merely random variations (Upton 1977). The loglinear model for the contingency table is expressed as follows:

Log (expected cell frequency) = Grand mean + Main effects parameters + Second and higher-order interactions. The grand mean is the average of the logs of frequencies in all table cells and is denoted by μ. The beauty of this model is to portray the multiple variables in multivariate situations where all the variables during the model run will interact with each other. The odds ratios of different factors were computed based on the second-order interaction parameters of the significant factors (Upton 1977).

The loglinear model used in this study is constructed from five-way contingency table of occupation, miners' age, experience, location, degree of injury. Degree of injury includes both injuries and no injuries.

The generalized form for the five-way contingency table shown is given as follows:

$$\begin{aligned}\mu_{abcde} = {} & \mu + \varphi_a^O + \varphi_b^A + \varphi_c^E + \varphi_d^L + \varphi_e^D + \varphi_{ab}^{OA} + \varphi_{ac}^{OE} + \varphi_{ad}^{OL} + \varphi_{ae}^{OD} + \varphi_{bc}^{AE} + \varphi_{bd}^{AL} + \varphi_{be}^{AD} + \varphi_{cd}^{EL} + \varphi_{ce}^{ED} \\ & + \varphi_{de}^{LD} + \varphi_{abc}^{OAE} + \varphi_{abd}^{OAL} + \varphi_{abe}^{OAD} + \varphi_{ace}^{OED} + \varphi_{acd}^{OEL} + \varphi_{ade}^{OLD} + \varphi_{bcd}^{AEL} + \varphi_{bce}^{AED} + \varphi_{bde}^{ALD} + \varphi_{cde}^{ELD} \\ & + \varphi_{abcd}^{OAEL} + \varphi_{abce}^{OAED} + \varphi_{abde}^{OALD} + \varphi_{aced}^{OEDL} + \varphi_{bcde}^{AELD} + \varphi_{abcde}^{OAELD}\end{aligned} \tag{7.1}$$

Constraints,

$$\begin{aligned}\sum_a \varphi_a^O = \sum_b \varphi_b^A = \ldots\ldots\ldots \sum_a \varphi_{ab}^{OA} = \sum_b \varphi_{ab}^{OA} = \ldots\ldots\ldots \sum_d \varphi_{abcde}^{OAELD} \\ = \sum_e \varphi_{abcde}^{OAELD} = 0\end{aligned} \tag{7.2}$$

where

μ_{abcde} = $\log_e(m_{abcde})$
m_{abcde} = expected cell frequencies.
φ_a^O = Main effects of the *a*th category of occupation variable
φ_b^A = Main effects of the *b*th category of age variable
φ_c^E = Main effects of the *c*th category of experience variable
φ_d^L = Main effects of the *d*th category of location variable
φ_e^D = Main effects of the *e*th category of degree of injury variable
φ_{ab}^{OA} = interaction between *a*th category of occupation and *b*th category of age.
φ_{ac}^{OE} = interaction between *a*th category of occupation and *c*th category of experience.
Similarly, we can describe the third-order interactions, fourth-order interactions, and fifth-order interactions as follows.
φ_{abc}^{OAE} = Third-order interaction between *a*th category of occupation, *b*th category of age and *c*th category of experience.
φ_{abcd}^{OAEL} = Fourth-order interaction between *a*th category of occupation, *b*th category of age, *c*th category of experience, and *d*th category of location.
φ_{abcde}^{OAELD} = Fifth-order interaction between *a*th category of occupation, *b*th category of age, *c*th category of experience, *d*th category of location, and *e*th category of degree of injury.

The occupational injury data were cross-classified by the variables of interest as shown in Table 7.1. To explore the univariate patterns of injuries, the crude risk ratios were computed with Statistical Packages for Social Sciences software application. This type of analysis may lead to very misleading results and is not sufficient to draw any conclusion.

7.2.3 Estimation of Parameters

The φ parameters of loglinear model are estimated by maximum likelihood estimation. The approach to maximum likelihood estimates is done by maximization of likelihood with respect to the model parameters. Computationally advantageous method includes maximum likelihood estimates of the expected frequencies, and from these, estimates of model parameters are computed. The observed cell frequencies (N_{abcde}) follow a multinomial distribution:

$$P(n_{11111},\ldots\ldots,n_{ABCDE}) = \left(\frac{n!}{\prod_{abcde} n_{abcde}!}\right) \times \prod_{abcde} (\pi_{abcde})^{n_{abcde}} \tag{7.3}$$

And the likelihood for the saturated loglinear is the natural logarithm of the observed cell frequencies,

Table 7.1 Cross-classification table for the variables

Sl. No.	Occupation	Age	Experience	Location	Degree of injury	
					Injury no	Injury
1.	Occu0	Age0	Exp0	Loc0	m_{00000}	m_{00001}
	…………	…………		Loc1	m_{00010}	m_{00011}
	…………	…………		Loc2	m_{00020}	m_{00021}
	…………	…………		Loc3	m_{00030}	m_{00031}
	…………	…………		Loc4	m_{00040}	m_{00041}
	…………	Age1	Exp1	Loc0	…………	…………
	…………	…………		Loc1	…………	…………
	…………	…………		Loc2	…………	…………
	…………	…………		Loc3	…………	…………
	…………	…………		Loc4	…………	…………
	…………	Age2	Exp2	Loc0	…………	…………
	…………	…………		Loc1	…………	…………
	…………	…………		Loc2	…………	…………
	…………	…………		Loc3	…………	…………
	…………	…………		Loc4	…………	…………
	…………	Age3	Exp3	Loc0	…………	…………
	…………	…………	…………	Loc1	…………	…………
	…………	…………	…………	Loc2	…………	…………
	…………	…………	…………	Loc3	…………	…………
	…………	…………	…………	Loc4	…………	…………
2.	Occu1	Age0	Exp1	Loc0	…………	…………
	…………	…………	…………	…………	…………	…………
	…………	…………	…………	…………	…………	…………
	…………	…………	…………	…………	…………	…………
	…………	…………	…………	…………	…………	…………
3.	Occu2	Age0	Exp1	Loc0	…………	…………
	…………	…………	…………	…………	…………	…………
	…………	…………	…………	…………	…………	…………
7.	Occu6	Age1	Exp1	Loc0	…………	…………
	…………	…………	…………	Loc1	…………	…………
	…………	…………	…………	Loc2	…………	…………
	…………	…………	…………	Loc3	…………	…………
	…………	…………	…………	Loc4	…………	…………
		Age3	Exp3	Loc0	m_{63300}	m_{63301}
				Loc1	m_{63310}	m_{63311}
				Loc2	m_{63320}	m_{63321}
				Loc3	m_{63330}	m_{63331}
				Loc4	m_{63340}	m_{63341}

$$\text{Ln(L)} = \text{Ln}\left(\left(\frac{n!}{\prod_{abcde} n_{abcde}!}\right) + \sum_{abcde} N_{abcde} \times \ln(m_{abcde}) - n \ln n\right) \quad (7.4)$$

since $\mu_{abcde} = log_e(m_{abcde})$

$$= \ln\left(\left(\frac{n!}{\prod_{abcde} n_{abcde}!}\right) + \sum_{abcde} N_{abcde} \times \mu_{abcde} - n \ln n\right) \quad (7.5)$$

where μ_{abcde} is a function of φ and μ.

In Eq. 7.5, first and third terms do not depend upon the parameters of the model, and thus, they are ignored in maximization of ln L. The log likelihood is maximized subject to the constraints in equation; estimation of the expected frequencies m_{abcde} in the saturated model of maximum likelihood estimates is given by the observed cell frequencies, which means $m_{abcde} = m_{abcde}$. Likelihood estimates of model parameters were computed by the following equations:

$$\mu_{abcde} = \ln m_{abcde} \quad (7.6)$$

$$\mu = \frac{\sum_{a,b,c,d,e} \mu_{abcde}}{ABCDE} \quad (7.7)$$

$$\mu_{a....=} \frac{\sum_{b,c,d,e} \mu_{abcde}}{BCDE} \quad (7.8)$$

$$\mu_{ab...=} \frac{\sum_{c,d,e} \mu_{abcde}}{CDE} \quad (7.9)$$

$$\mu_{abc..=} \frac{\sum_{d,e} \mu_{abcde}}{DE} \quad (7.10)$$

$$\mu_{abcd.=} \frac{\sum_{e} \mu_{abcde}}{E} \quad (7.11)$$

$$\varphi_a^O = \mu_{a....} - \mu \quad (7.12)$$

$$\varphi_e^D = \mu_{abcd.} - \mu \quad (7.13)$$

$$\varphi_{ab}^{OA} = \mu_{ab...} - \mu_{a....} - \mu_{.b...} + \mu \quad (7.14)$$

$$\varphi_{bc}^{AE} = \mu_{bc...} - \mu_{.b...} - \mu_{..c..} + \mu \quad (7.15)$$

$$\varphi_{abc}^{OAE} = \mu_{abc..} - \mu_{ab...} - \mu_{.bc.} - \mu_{a.c..} + \mu_{a....} + \mu_{.b...} + \mu_{..c..} - \mu \quad (7.16)$$

$$\varphi_{bcd}^{AEL} = \mu_{bcd} - \mu_{.b.d.} - \mu_{.bc..} - \mu_{..cd.} + \mu_{d....} + \mu_{.b...} + \mu_{..c..} - \mu \quad (7.17)$$

The φ parameters of the saturated loglinear model are estimated by means of an iterative proportional fitting algorithm. To test the null hypothesis that the parameter φ values are 0 can be based on Z value, where Z is the ratio of parameter estimate to its standard error, since the standardized φ is approximately normally distributed with a mean of 0 and standard deviation of 1 if the model fits the data. φs with Z values greater than 1.96 in absolute value can be considered significant at the 0.05 probability level. The goodness-of-fit tests for general loglinear model are the following. The test of hypothesis that a particular model fits the observed data can be based on the Pearson chi-square statistics which is calculated by the following formula:

$$\chi^2 = \sum_{i,j} \frac{(O_{ij} - E_{ij})}{E_{ij}} \tag{7.18}$$

where

O_{ij} = observed frequency in ith row and the jth column of the contingency table.
E_{ij} = expected frequency in ith row and the jth column of the contingency table.

An alternative goodness-of-fit static for loglinear model is the likelihood ratio chi-square, which is calculated by the following formula.

$$\mathrm{L}^2 = 2 \times \sum_{i,j} F_{ij} \times \log \frac{F_{ij}}{F'_{ij}} \tag{7.19}$$

F_{ij} = observed frequency in ith row and the jth column of the contingency table.
F'_{ij} = expected frequency in ith row and the jth column of the contingency table.

The advantage of likelihood ratio test is that it, however, has important properties that χ^2 does not possess, it is static that is minimized by MLE, and like the total sum of squares in analysis of variance, it can conveniently be broken into interpretable parts that add up to the total. They have an asymptotic chi-square distribution with degrees of freedom equal to the number of independent parameters in the model. Lack of fit of model is indicated by the large values of these statistics corresponding to small probability values. The saturated model fits perfectly to the data as it reproduces the data exactly.

7.2.4 Data Collection

The injury experience and no injury data have been collected from log books and registers of an underground hard rock mine from 2002 to 2013 which is maintained by Safety and Personnel Department in Form 'J' and form 'K'. The mine is located in the eastern part of India, and from mining operation point of view, these are the most important production areas in India which is chosen for the case study and operates six days in a week and three shifts per day A, B, and C, respectively.

Table 7.2 Geological and mining information

Entry into the mine	8 degree decline
The host rocks for mineralization	feldsphatic-schist, chlorite sericite schist with magnetite, apatite and quartzite
Number of lodes	Six lodes with wide variation in shape, size, and continuity
Method of mining	Horizontal cut and fill
Manpower	607

The data contained 44 reportable injuries and 14 non-reportable injuries, and rest are no injuries. The geological and mining information of the mine are shown in Table 7.2.

7.3 Results and Analysis

Table 7.3 shows the man-days worked during the years 2002–2013.

Table 7.4 shows the number of man-days lost and the man-days lost per injury in each occupation group. The man-days lost by casual and regular workers are more because there is a fatal accident has occurred in this group. Helper's occupational group restrains more number of man-days lost after casual and regular workers, and simultaneously, drill man has slightest man-days lost.

Table 7.5 shows the cross-classification of occupation and cause of accidents because the slip and fall are more compared to others. Miners with occupation casual and regular are prone to slip and fall accidents more (10), helpers are getting injured more due to fall of roof (6), slip and fall (5), caught in between objects (4). Accidents due to suffocation by gas and electric shock are too less in comparison with other causes of accidents. Also, operators and drill men are experiencing less accident than others.

Table 7.3 Man-days worked during the years 2002 to 2013

Year	Man-days worked per year
2002	113,453
2003	64,707
2004	71,398
2005	82,498
2006	111,848
2007	13,884
2008	120,418
2009	30,283
2010	183,854
2011	157,639
2012	146,617
2013	140,137

Table 7.4 The number of man-days lost and the man-days lost per injury in each occupation group

Occupation	Frequency	Percentage	Man-days lost			Man-days lost per injury
			Minimum	Maximum	Total	
Helper	20	34.48	2	269	485	24.25
Operators	6	10.34	1	22	37	6.17
Tradesman	6	10.34	3	48	118	19.67
Drill man	2	3.45	1	4	5	2.5
Casual and regular workers	19	32.76	1	6000	6404	337.05
Others	5	8.62	2	33	59	11.80

7.3.1 *Injury Rate Analysis*

In injury rate analysis, the normalized injury rates are calculated with respect to the variables and categories. The frequencies of individual, workplace variables are given in the table. Normalized injury rates are calculated using the following formula:

$$\text{NIR for category j} = \frac{\frac{\%\textit{injured in category j}}{\%\textit{population in category j}}}{\sum_{i=1}^{n} \frac{\%\textit{injured in category i}}{\%\textit{population in category i}}} \times 100 \qquad (7.20)$$

From Table 7.6 and Table 7.7, normalized injury rates (NIR) for all injuries show that the age group (Age1) between 40 and 50 years is having more injury rate compared to the remaining groups, and reportable injury rates (42.49) are high in this age group. Non-reportable injury rates are high in the age group (Age0) in between 20 and 30 years as compared to remaining groups, and reportable injuries are less. For the age group of 30 to 40 (Age1), the injury rates are nearly same for all total injuries and non-reportable and reportable injuries. The age group more than 50 year (Age3) has less injury rates in total and non-reportable injuries. The normalized injury rates for experience variable show that miners having experience between 10 and 20 years have higher total injury and non-reportable and reportable injury rates. Miners of more than 30 years of experience are having zero injury rates for total injury and non-reportable and reportable injuries. Injury rates for experience workers (Exp0) between 10 and 20 years are nearly equal for all, i.e. total injury and non-reportable and reportable injuries. Occupation group of casual and regular workers (Occu1) are having highest injury rates for all injuries (28.53) and reportable injuries (49.17). Operators of casual and regular (Occu5) are having highest non-reportable injury rate (44.13) which is compared to remaining occupations. Blasters are not injured in between 2002 and 2013 so they are having zero injury and non-reportable and reportable injuries. Helper's injury rate for all injury

Table 7.5 Cross-classification of occupation and cause of accident

Cause	Fall of object	Fall of roof	Fall of side	Caught in between objects	Slip and fall	Struck by objects	Flying pieces and Foreign body	Suffocation by gas	Electric Shocks	Others
Occupation										
Helper	1	**6**	0	**4**	**5**	2	1	1	0	0
Operators	0	2	0	0	1	0	1	0	1	1
Tradesman	1	0	0	1	2	0	0	0	0	2
Drill man	0	0	1	0	0	0	1	0	0	0
Casual and regular workers	3	2	1	0	**10**	2	1	0	0	0
Others	0	0	0	0	3	2	0	0	0	0

Table 7.6 Distribution of injuries for total, non-reportable, reportable with respect to age, experience, occupation, and workplace location

Degree of injury	All injury		Non-reportable		Reportable		Total manpower population	
Variable category	Frequency	Percentage	Frequency	Percentage	Frequency	Percentage	Frequency	Percentage
Individual variable								
Age								
Age0 20–30	18	31.03	8	57.14	10	22.73	269	44.32
Age1 30–40	19	32.76	3	21.43	16	36.36	182	29.98
Age2 40–50	19	32.76	3	21.43	16	36.36	120	19.77
Age3 > 50	2	3.45	0	0.00	2	4.55	36	5.93
Experience								
Exp0 1–10	34	58.62	10	71.43	24	54.55	454	74.79
Exp1 10–20	22	37.93	4	28.57	18	40.90	74	12.19
Exp2 20–30	2	3.45	0	0	2	4.55	77	12.69
Exp3 > 30	0	0	0	0	0	0	2	0.33
Occupation								
Occu0 Helper	20	34.48	4	28.57	16	36.36	386	63.96
Occu1 Operators	6	10.34	4	28.57	2	4.55	28	4.610
Occu2 Tradesman	6	10.34	1	7.14	5	11.36	54	8.90
Occu3 Blaster	0	0	0	0	0	0	19	3.130
Occu4 Drill man	2	3.46	1	7.14	1	2.27	33	5.44
Occu5 Casual & Regular workers	19	32.76	3	21.43	16	36.36	56	9.03
Occu6 Others	5	8.62	1	7.14	4	9.09	31	5.11

(continued)

Table 7.6 (continued)

Degree of injury	All injury		Non-reportable		Reportable		Total manpower population	
Workplace variable								
Workplace location								
Loc0 Haulage	8	13.79	1	7.14	7	15.91	57	9.39
Loc1 Development place	24	41.38	7	50.0	17	38.64	164	27.02
Loc2 S/f and Workshops	14	24.14	3	21.43	11	25	72	11.86
Loc 3 Raise	1	1.72	0	0	1	2.27	74	12.19
Loc4 Stope	11	18.97	3	21.43	8	18.18	240	39.54

Table 7.7 Normalized injury rates for total injury, non-reportable, reportable with respect to age, experience, occupation, workplace location

Variable category	Degree of injury		
	Total injury	Non-reportable	Reportable
Individual variable			
Age			
Age0 20–30	17.37	41.88	11.78
Age1 30–40	27.05	23.05	27.94
Age2 40–50	41.19	35.06	42.49
Age3 > 50	14.39	0	17.78
Experience			
Exp0 1–10	18.74	29.09	16.40
Exp1 10–20	74.76	70.91	75.51
Exp2 20–30	6.49	0	8.09
Exp3 > 30	0	0	0
Occupation			
Occu0 Helper	7.34	3.57	6.28
Occu1 Operators	30.03	49.17	10.92
Occu2 Tradesman	15.76	6.98	14.11
Occu3 Blaster	0	0	0
Occu4 Drill man	5.03	10.39	4.63
Occu5 Casual and regular workers	28.53	18.79	44.43
Occu6 others	13.32	11.10	19.63
Workplace location			
Loc0 Haulage	25.97	15.32	28.74
Loc1 Development face	27.03	37.3	24.32
Loc2 Surface (s/f) and workshops	36.04	36.49	35.88
Loc3 Raise	2.47	0	3.24
Loc4 Stope	8.48	10.89	7.82

and non-reportable injuries are low compared to remaining groups. Drill man experiences lower injury rate for non-reportable. Occu6 (others) having nearly equal injury and non-reportable and reportable injuries. In location variable, category surface and workshop (Loc2) are having highest percentage of injury rate for all injury and reportable injuries. Development face (Loc1) is having higher injury rate for non-reportable injuries. Raise (Loc3) is having lowest injury rate for non-reportable, reportable, and total injuries. After Raise, stope (Loc4) has to restrain lowest number of injury rate for total injury, non-reportable, reportable injuries.

Table 7.8 K-way and higher-order effects of loglinear model

K		Degree of freedom (df)	Likelihood ratio		Pearson	
			Chi-square	Significance	Chi-square	Significance
K-way and Higher-order effects	5	216	1.202	1.000	0.647	1.000
	4	666	44.245	1.000	37.062	1.000
	3	993	286.129	1.000	2095.121	0.000
	2	1102	1310.718	0.000	8022.372	0.000
	1	1119	3745.142	0.000	33352.802	0.000

7.3.2 Loglinear Modelling

The main effects represent the increments or decrements from the base value (μ) for particular combination of the values of row and column variables. Positive values of main effects occur when the average number of cases in a row and column is greater than the overall average. The interaction parameters show how much difference is there in between the sums of the effects of the variables taken collectively and individually. The Statistical Packages for Social Sciences higloglinear routine was used for the analysis of loglinear model. Salient features of the model run are discussed in this section. The effects of the main and interactions are shown in Table 7.8. The first line of table is a test of hypothesis that the fifth-order interaction (K = 5) is zero. This is the goodness-of-fit statistical model without fifth-order interaction. Similarly, the entry for the K of 4 is the goodness-of-fit static for a model without fourth- and fifth-order interactions. The last line (K = 1) corresponds to a model that has no effects except grand mean (μ). That is, the expected value for all the cells is the same and equal to the average of the logs of observed frequencies in all cells. Small observed significance level indicates that the hypothesis of particular orders with zero should be rejected. Since the significance level for the test that fourth- and fifth-order terms are 0 and 1, the hypothesis for the fourth and fifth order zero should not be rejected. Thus, it appears that a model up to third order is enough to explain variations in observed cell frequencies. The estimated parameters (φs) of the loglinear model with their Z values are shown in Table 7.10. The φs with Z values greater than 1.96 can be considered significant at the 0.05 level. From Table 7.10, it is clear that none of the third-order interactions except Occupation*Age*Experience, Occupation*Experience*DOI, and Occupation*Location*DOI are significant. The contribution of third-order interactions is very negligible in explaining the variations in observed cells. Table 7.8 also shows that the tests k-way and higher-order effects in loglinear model are zero.

Table 7.9 shows the partial associations between all factors up to their second-order interactions. The partial associations between the factors show their contribution towards the explaining variation between observed cell counts of the contingency table. Table 7.9 proves that the main effects of the age, experience, location, and occupation and their associations with DOI (Degree of injury) are significant (Except Age*DOI). Since the second-order terms of the partial associations shows that the factors Experience, location, and occupation are related to the

Table 7.9 Associations between all factors up to their second-order interactions

Effect	Degree of freedom	Partial Chi-square	Significance*
Occupation*Age	18	21.251	0.267
Occupation*Experience	18	21.135	0.273
Age*Experience	9	373.265	**0.000***
Occupation*Location	24	323.891	**0.000***
Age*Location	12	18.974	0.089
Experience*Location	12	8.058	0.781
Occupation*DOI	6	42.482	**0.000***
Age*DOI	3	1.113	0.774
Experience*DOI	3	25.928	**0.000***
Location*DOI	4	19.300	**0.001***
Occupation	6	804.106	**0.000***
Age	3	213.413	**0.000***
Experience	3	766.960	**0.000***
Location.	4	191.116	**0.000***
DOI	1	458.830	**0.000***

* *Significance is less than 0.05 levels*

DOI (Degree of injury), Excluding Age*DOI (Degree of injury) further, among remaining four factors there are associations between them, particularly between Age*Experience and Occupation*Location. Hence, it is concluded that all of the four factors mentioned age, experience, occupation, and location are not completely independent.

7.3.3 Risk Estimation Using Loglinear Model

The results of the loglinear model have shown that the variations observed in the cell frequencies of the contingency table are not random. The factors of experience, occupation, and location are significantly influencing the injuries. In the output of the loglinear model in SPSS, we get the required parameter estimation. The parameter estimates along with their significance level are shown in Table 7.10. These parameter estimates can be labelled as "b" coefficients, and the Exp (b) is the odds ratio. Odds ratio 1 indicates no effect, greater than 1 indicating the variable in question increases the odds, and less than 1 indicating the variables decreases the odds ratio (Garson 2008). The values obtained from SPSS program used to calculate the odds ratio and parameter estimates with significance level 0.05 are taken into consideration. The effects and parameter estimates (b) are shown in Table 7.10. The odds ratio (Exp (b)) is high, i.e. 1.883 times for the effect Occupation*Age*Experience, and low, i.e. 1.153 times for effect Experience* DOI

Table 7.10 The effects and parameter estimation in loglinear model

Effect	Parameter number	Estimates	Standard Error	Z > 1.96	Significance < 0.05	Odds ratio (Exp(b))
Occupation*Age*Experience	1	0.633	0.229	2.763	0.006	1.883
Occupation*Experience*DOI	1	0.321	0.146	2.206	0.027	1.379
Occupation*Location*DOI	11	0.451	0.186	2.422	0.015	1.57
Occupation*Experience	1	0.444	0.146	3.048	0.002	1.559
Age*Experience	1	0.406	0.112	3.636	0.000	1.501
Occupation*Location	11	0.484	0.186	2.602	0.009	1.623
Occupation*DOI	1	0.276	0.092	3.006	0.003	1.318
Experience*DOI	1	0.142	0.067	2.124	0.034	1.153

Table 7.11 Odds ratio of second-order interaction terms for loglinear model

Effect	Parameter number	Estimates	Standard error	Z > 1.96	Significance < 0.05	Exp(b)
Age*Experience	1	0.406	0.112	3.636	0.000	1.501
Occupation*Location	11	0.484	0.186	2.602	0.009	1.623
Occupation*DOI	1	0.276	0.092	3.006	0.003	1.318
Experience*DOI	1	0.142	0.067	2.124	0.034	1.153

There are no fourth-order interaction terms that are significant. So, we cannot find the odds ratio. After evaluation of third-order interactions, the second-order interactions were evaluated and important results are given in Table 7.11. The odds ratios of second-order interaction in terms of loglinear model are evaluated in the following table. The odds ratio shown in this table reveals that the workers with age group 20 to 30 years and having experience between 1 and 10 years are more susceptible to face injury than other group. This is statistically significant which is reflected in the table. They are 1.623 and 1.153 times, respectively, more risky than other group during working in the case study mine.

7.3.4 *Discussion*

In this research, the injury rates were examined with respect to the variables and categories. The injury rates are calculated based on the normalized injury rate (NIR). The advantage of this method is that this method considers population figure, thus giving weighted pattern of injuries that were studied. For the case study in hard rock mine, the injury rate shows that increase in age results increases injury rate, and concurrently, location variable shows the development faces and the surface and workshops are having high injury rates.

Loglinear model proves that the variations observed in cell frequencies are not random. The variables occupation, workplace location, experience have significant effect on observed injury frequencies. The main and second-order interactions of variables sufficiently explained variations in injury frequencies. The experience category shows that the miners of experience with more than 30 years show the less injury rate and miners of experience with 1 to 10 years show more injury rate, whereas the occupation variable shows the casual and regular workers have more injury rate. From the output of the loglinear model by SPSS, we get the estimated parameters. The odds ratio of the different parameter has been calculated from the estimated parameters. Workers with age group in between 20 and 30 years and having experience between 1 and 10 years are more susceptible to face injury than other groups which is statistically significant. They are 1.623 and 1.153 times, respectively, more risky than other group during working in case study mine.

7.4 Conclusions

The findings of the loglinear model analysis indicate the factors affect on the occurrence of an injury in a hard rock mine. The reason may be justified as younger and inexperience or less experience mine workers are taking more risk during their work in mine in compared to the aged and more experienced group. However, the young group of workers would have been allotted cooperative heavy job in regular basis, and they also always try to complete their job as fast as possible by taking more risk and faced injury frequently. Identification of these risk factors of injuries will provide valuable information in injury-preventive programmes. Specifically, the management should pay due attention towards the problems of working environment and safety of the workers.

7.5 Limitation of the Study

1. The study has been conducted only in single underground hard rock mine. So, "mine" could not be taken as a variable of interest in this study which might have tremendous impact in the model.
2. The very important statistical models, namely logit and logistic models, within the scope of loglinear model have not been done in this study. The perfect categorical risk of considered variable of interest could be estimated in different multivariate situations if these above models were established.
3. In this loglinear model, only linear relationships between the variables of interest have been assessed. Any nonlinearity between the variables of interest has not been tested.

References

Bhattacherjee, A., Kunar, B.M.: Epidemiology: a tool for risk analysis of mine injuries. Transactions of the Society of Mining, Metallurgy, and Engineering (SME) USA, vol. 328, pp. 490–500 (2011)

Bhattacherjee, A., Kunar, B.M., Chau, N.: Contribution of physical hazards and health-related factors in occupational injuries of underground coal miners: a matched case-control study. Proceedings of the 32nd International Conference of Safety in Mines Research Institutes (SIMRI), Beijing, China, September 28–29, SAWS, China, pp. 423–430 (2007)

Garson, D.G.: Factor Analysis: Statnotes. Retrieved March 22, 2008, from North Carolina State University Public Administration Program (2008),http://www2.chass.ncsu.edu/garson/pa765/factor.htm

Ghosh, A.K.: Risk assessment of occupational injuries in underground coal mines. J. Min. Met. Fuels **58**(9), 243–248 (2010)

Ghosh, A.K., Bhattacherjee, A., Chau, N.: Relationships of working conditions and Individual characteristics to occupational injuries: a case control study in coal miners. J. Occup. Health **46**, 470–478 (2004)

Ghosh, A.K., Bhattacherjee, A.: Predictors of occupational injuries among coal miners: a causal analysis. Min. Technol. (Transaction Institute of Materials, Minerals and Mining, United Kingdom and Australian Institute of Mining and Metallurgy), **116**(1), 16–24 (2007)

Ghosh, A.K., Bhattacherjee, A.: Loglinear model for assessment of risk factors of occupational injuries in underground coal mines. J. Geol. Min. Res. **1**(2), 25–33 (2009)

Kunar, B.M., Kumar, P. P.: Risk assessment and safety evaluation in underground metal mines. Proceedings of the 2nd National Seminar on "Underground Metal Mining: Status and Prospects" hosted by Indian School of Mines, Dhanbad, 13–15th October, 2011, pp. 329–335 (2011)

Maiti, J., Bhattacherjee, A.: Evaluation of risk of occupational injuries among underground coal mine workers through multinomial logit analysis. J. Saf. Res. **30**(2), 93–101 (1999)

Maiti, J., Bhattacharjee, A., Bangdiwala, S.I.: Loglinear model for analysis of cross tabulated coal mine injury data. Inj. Control Saf. Promot. **8**(4), 229–236 (2001)

National Research Council/National Academy of Sciences Towards safer underground coal Mines. Material Academics Press, Washington, DC (1982)

Onder, M., Adiguzel, E.: Evaluation of occupational fatalities among underground coal mine workers through hierarchical loglinear models. Ind. Health **48**, 872–878 (2010)

Onder, M., Onder, S., Adiguzel, E.: Applying Hierarchical Loglinear Models to Nonfatal Underground Coal Mine Accidents for Safety Management. Int. J. Occup. Saf. Ergon. **20**(2), 239–248 (2014)

Pal, B.K., Khanda, D.K.: Risk management-an effective tool for accident prevention in mining industry. In: Proceedings of National Seminar on Rock Machine Interaction in Excavations, Varanasi, March 7–8, pp. 183–193 (2008)

Rahman, N.A.A., Ismail, A.R., Mansor, M.A.: Analysis of the perception of occupational accident in mining and quarry sector towards safe and healthy working environment. Int. J. Curr. Res. Acad. Rev. **1**, 95–102 (2014). ISSN 2347–3215

Upton, J.G.: The analysis of cross-tabulated data. Wiley, New York (1977)

Chapter 8
Mathematical Modelling of Human Energy Consumption During Hand Arm Vibration in Drilling Operation for Female Operator

Chandrashekhar D. Mohod and Ashish M. Mahalle

Abstract Vibration transmission in the hand of operators while working with handheld power tools is one of the important concerns with respect to the safety of operators. The handheld power tools are widely used in manufacturing industry, construction industry, service centres, road construction, etc. The operation on powered handheld tools such as grinders, drills, etc., exposes workers to hand arm vibration. Hand arm vibration (HAV) is the vibration transmitted to the parts of hand arm during the operation of handheld power tools. Vibrations transmitted can originate several types of illness with associated symptoms of blood supply, nerves, muscles of hand arm system and lead neurological, vascular and oesteoarticular disorders. Vibration-induced white finger is one of the vibration-induced disorders in fingers and hand because of blanching along with tingling and numbness in fingers and hand. Workers who are repeatedly exposed to intense hand-transmitted vibration are at risk of developing health problems. The research is focused to analyse effects of hand arm vibration on human energy consumption during drilling operation by female operators. For experimentation, seven different postures and two age group of operator between 18 and 40 years were identified. A mathematical model is developed for human energy consumption by female operator considering the different parameters, viz. age, weight, posture angle of operator during working, hardness of commonly used different materials and push force applied by the operator. Dimensional analysis is used for developing the mathematical model. Actual human energy consumption is calculated considering the pulse rate, age and weight of operator, etc. Artificial neural network (ANN) is developed for human energy consumption using MATLAB software. Then actual human energy consumption is validated with mathematical model and ANN prediction model. Sensitivity analysis is used to identify the most dominant parameter in mathematical

C.D. Mohod (✉) · A.M. Mahalle
Laxminarayan Institute of Technology, Nagpur, MS, India
e-mail: cdmohod@gmail.com

J. Maiti and P.K. Ray (eds.), *Industrial Safety Management*,
Managing the Asian Century, https://doi.org/10.1007/978-981-10-6328-2_8

model for human energy consumption by operators. Age of the operator is observed to be a most dominant factor in human energy consumption. Mathematical modelling, ANN and experimental results are well agreed with R2 value 0.99 and absolute error of 1.1898.

Keywords Human energy · Hand arm vibration · Working posture ANN

8.1 Introduction

In today's era of industrialization, atomization is adopted in maximum fields. While performing the operations, operators use various handheld power tools. The handheld power tools are widely used in various areas such as manufacturing industry, construction sites, carpenter industry, workshop, automobile service centre, road construction, etc. While working with handheld power tools, vibration generated by these tools is transmitted to the hands of the operator. Transmission of vibration by hand arm system (HAS) causes some ill effects, and continuous working in such conditions may leads to dangerous injuries or permanent damage of hand arm organs. For the handheld power tool, tool characteristics, working material, hand-handle contact area, hand-handle coupling force, working posture, work organization, and various subjective characteristics affect transmission of vibration. It also depends on age and gender of operator, operator skill, Individual habits, medical history, etc. Early signs of hand arm vibration syndromes are—tingling, pain and numbness in the fingers, loss of feeling in the fingers, weakness in the hands. While performing any operation, human energy gets utilized. Consumption of human energy depends on pulse rate or blood pressure of operator, age and weight of operator. Human energy consumption can be the parameter for study of vibration transmission in hand arm.

8.2 Literature Review

Vibration transmitted to hand arm system due to handheld tools is one of the major aspects in designing handheld tools. The hand-handle contact force is dependent on grip force, push force and handle diameter. Contact force has linear relation with handle diameter (Gerhardsson and Balogh 2005). Total amount of absorbed power by hand arm system would likely to be most dependent on push force (Dowell et al. 2006). Higher power absorption is observed in extended arm posture than flexed arm posture along Z-axis (Neely and Burstorm 2006). In actual practice, hand tools are not necessarily in said position only and hence there is the need to analyse other different postures with extended and flexed both in high and low positions (Neely and Burstorm 2006), Dong and Ren (2008). Transmission of vibration is dependent upon various factors such as hand arm posture, vibration direction, grip and push force, handle diameter, magnitude of vibration, type of gender, etc. (Dowell et al. 2006). The research shows gender difference gives different responses to hand arm vibration. On average rating of perceived intensity and discomfort was higher for female than male by few investigated frequencies (Dong and Welcome 2006). In survey, 12.2% of female workers had experienced difficulties associated with vibration in their work environment (Dong and Welcome 2006). Through the literature review, the gap in existing research is identified. Age of operators and thus energy consumption has been considered here as a scope for further research.

8.3 Operational Methodology

The following subjects/conditions are considered:

- Operators—The two age groups of five each female operators 18–25 years and 25–40 years have been identified.
- Working Surfaces—Table 8.1 represents most commonly used selected materials with hardness for drilling operation.
- Arm posture—Seven postures, viz. 0°, 45°, 90°, 135°, 180° (extended arm) and 90°–90° (225°), 90°–180° (270°) (flexed arm) with elbow angle 90° are identified.
- Position on hand arm Vibration is measured in terms of velocity and acceleration at four different locations.

The different parameters measured are

1. Velocity and acceleration measurement,
2. Anthropometric data of operator, and
3. Pulse rate of operator.

The drilling operation is performed on all different material in all postures for female operators of all age group. The working environment was at ambient temperature 28°–32 °C. Before the start of the drilling operation, pulse rate and oxygen consumption are measured.

Table 8.1 Surface material with hardness

Material	Mild Steel	Concrete	Aluminium	Wood
Hardness (BHN*)	60	40	20	10

**BHN* Brinell hardness number

8.4 Mathematical Modelling

8.4.1 Development of Mathematical Model

There are many factors affecting the performance of drilling operation. To study drilling operation over different material of different hardness, field data modelling approach is proposed.

Identification of dependent and independent variables of the phenomenon is taken based on known qualitative physics of the phenomenon. After getting the experimental results, adopting the appropriate method for test data checking and rejection, the erroneous data is identified and removed from the gathered data. Based on this purified data, quantitative relationship between the dependent and independent in terms of π is defined.

8.4.2 Identification of Variables or Quantities

The first step in this process is identification of variables. The dependent or the response variable, in this case, is consumption of Human Energy (HE) in drilling operation. The different independent variables, which are involved, have been grouped in the following categories.

- Anthropometric features/variables of operator
- Drill rod parameters
- Drilling machine/Process parameters
- Other variables

Dimensional analysis is used for reduction of independent variables. The exact mathematical form of this dimensional equation is the targeted model.

8.4.3 Drilling Operation (Independent Variables) Parameters

Identification of dependent and independent variables of the phenomenon is to be done based on known qualitative physics of the phenomenon. The drilling phenomenon is influenced by variables as mentioned in Table 8.2.

Table 8.2 Independent and dependent variables in drilling operation

Sr. No.	Description	Variables	Symbol	Dimension
1	Age of the operator (Ag)	Independent	Ag	$[M^0L^0T^0]$
2	Weight of the operator (W)	Independent	W	$[M^1L^0T^0]$
3	Posture adopted by operator (Po)	Independent	Po	$[M^0L^0T^0]$
4	Hardness of material to be drill (H)	Independent	H	$[M^0L^0T^0]$
5	Force	Independent	Pf	$[M^1L^0T^0]$
6	Diameter of drill (Dd)	Independent	Dd	$[M^0L^1T^0]$
7	Length of Drill (Dl)	Independent	Dl	$[M^0L^1T^0]$
8	Speed of machine (Nd)	Independent	Nd	$[M^0L^1T^{-1}]$
9	Human energy (HE)	Dependent	HE	$[M^1L^2T^{-2}]$

8.4.4 Formulation of Field Data-Based Model

Five independent π terms ($\pi_1, \pi_2, \pi_3, \pi_4, \pi_5$) and one dependent π term (Z1) have been identified for field study model formulation.

Each dependent π term is a function of the available independent π terms,

$$Z_1 = f(\pi_1, \pi_2, \pi_3, \pi_4, \pi_5), \tag{8.1}$$

where

$$\begin{aligned} Z_1 &= \pi_{D1}, \text{First dependent } \pi \text{ term} \\ &= HE/(W * Vd^2) \\ (Z) &= K\{[Ag]^a[Po]^b[H]^c[Pf/W]^d[Dd/Dl]^e \end{aligned} \tag{8.2}$$

8.4.5 Mathematical Model for Female Operator

$$\begin{pmatrix} Z \\ ZA \\ ZB \\ ZC \\ ZD \\ ZE \end{pmatrix} = \begin{pmatrix} n & A & B & C & D & E \\ A & A^2 & AB & AC & AD & AE \\ B & AB & B^2 & BC & BD & BE \\ C & AC & BC & C^2 & CD & CE \\ D & AD & BD & CD & D^2 & DE \\ E & AE & BE & CE & DE & E^2 \end{pmatrix} \begin{pmatrix} k \\ a \\ b \\ c \\ d \\ e \end{pmatrix}$$

$$(Z_1) = 1.0981 * (\pi_1)^{0.5971} * (\pi_2)^{0.0031} * (\pi_3)^{0.0128} * (\pi_4)^{0.0403} * (\pi_5)^{-0.0815}. \tag{8.3}$$

8.5 Artificial Neural Network Simulation

8.5.1 Human Energy by Artificial Neural Network (ANN)

In complex phenomenon involving nonlinear kinematics where in the validation of field data-based models is not in close proximity; it becomes necessary to formulate Artificial neural network (ANN) simulation of the observed data. To validate the results of human energy consumption calculated using developed mathematical equation and from field values, ANN prediction model is used. MATLAB software is used for developing ANN simulation.

8.5.2 ANN Model Development

Human energy consumption data for female work is trained for ANN simulation with performance level 0.400529 compared to goal 0.99 is shown in Fig. 8.1.

The error in the calculated and predicted human energy consumption for female operator by ANN is represented in Fig. 8.2. Experimental data is trained for prediction of human energy consumption. Figure 8.3 represents training of data for ANN simulation with performance level matching to goal level value zero.

Validation of training and testing data for human energy consumption is represented in Fig. 8.4. The regression analysis for human energy consumption by female operators of actual and computed data by ANN is represented in Fig. 8.5. From the conventional approach and ANN simulation, the curve obtained by dependent terms for drilling operation human energy D is overlapping due to the less percentage of error which gives an accurate relationship between ANN simulation and experimental data.

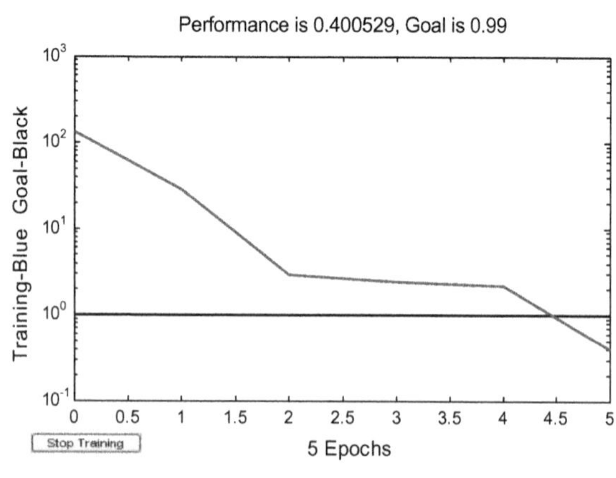

Fig. 8.1 Training of the network for HE

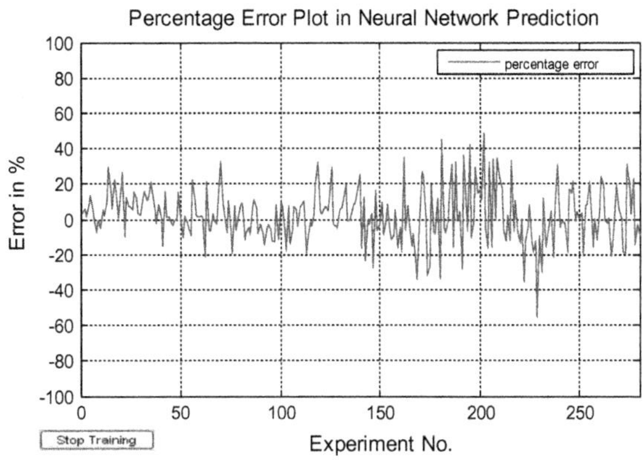

Fig. 8.2 % error plot prediction of HE

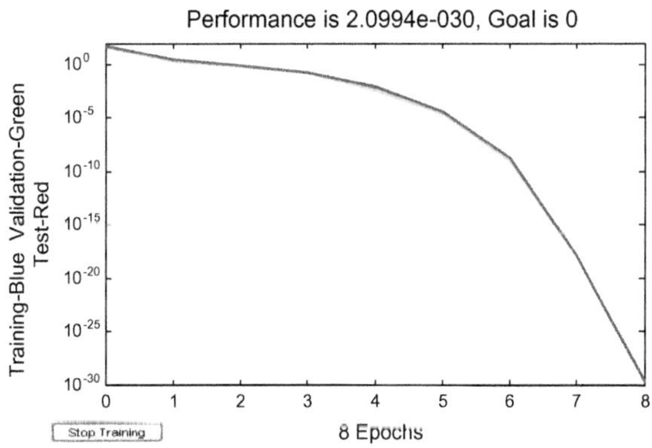

Fig. 8.3 Training of experimental data for HE

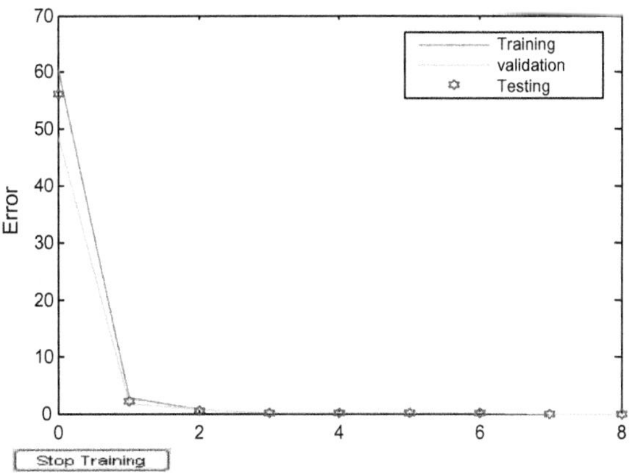

Fig. 8.4 Training, validation, testing of experimental data

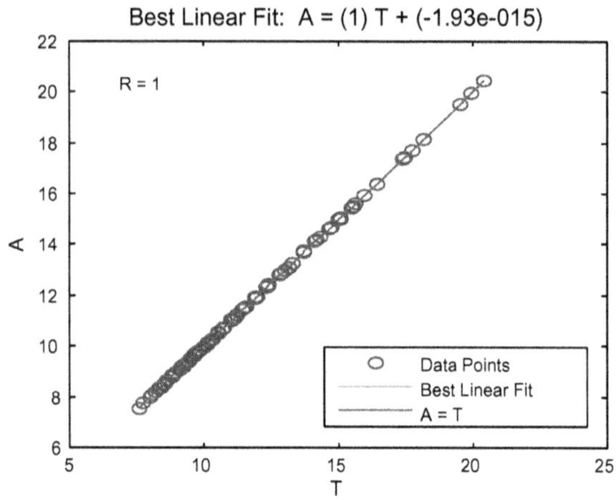

Fig. 8.5 Regression analysis comparisons of actual and computed data by ANN

8.5.3 *Human Energy Calculation from Field Value*

The human energy consumption by the operator is calculated.

$$\textit{Human Energy Expenditutre} = (0.4475 \times \textit{pulse rate}) + (0.1263 \times \textit{weight}) + (0.074 \times \textit{age}). \tag{8.4}$$

Values of human energy expenditure depend on change in pulse rate, age and weight of operator. The calculated values are used for predicting the human energy consumption by ANN simulation.

8.5.4 *Sensitivity Analysis of Drilling Operation*

Through the technique of sensitivity analysis, the change in the value of a dependent π term as a result of changing in the value of individual independent π term is evaluated. The average value of the change in the dependent terms due to the introduced change of 20% in each term is calculated. Table 8.3 represents the summarized data. Figure 8.6 is the graphical representation of variation in π terms. In the consumption of human energy by female operator π1, i.e., age of operator is more dominant π term.

Table 8.3 Percentage change in Z1 of each π term

Sr. No	Term	Z_1 (%)
1	π_1	11.95
2	π_2	0.06
3	π_3	0.26
4	π_4	0.81
5	π_5	−1.64

8.6 Results and Discussion

Human energy consumption is evaluated by three different methods, viz. mathematical model, field value and ANN prediction model. Figures 8.7, 8.8 and 8.9 represent the human energy obtained by mathematical model, field values and ANN model. Ten female operators worked at seven posture angles, and human energy consumption for 70 experiments performed on material, viz. mild steel, concrete, aluminium and wood is obtained. Figure 8.7 represents human energy consumed by female operator obtained by experimentation and ANN predicted model. Human energy consumption is more as the age of operator increases. From Fig. 8.7, it is clear that human energy consumption predicted by ANN model is matching with human energy values obtained from the experimental equation because ANN model has very negligible error while training and predicting the values.

Human energy consumption obtained by experimentation equation and mathematical model equation is represented in Fig. 8.8. Human energy consumption obtained from mathematical equation (blue colour) is less than calculated by experimental results (red colour). Age of operator is more dominant factor in mathematical equation while in field value calculation pulse rate change is considered more dominant factor and hence both values are different.

Figure 8.9 represents human energy consumption by female operator obtained by three different methods. Human energy consumption predicted by ANN model and obtained by experimental results is nearly same while obtained from

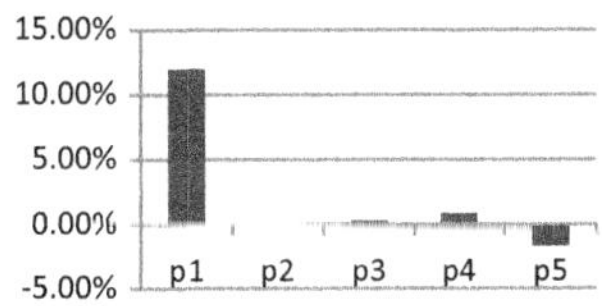

Fig. 8.6 % Change in HE with variation of π term

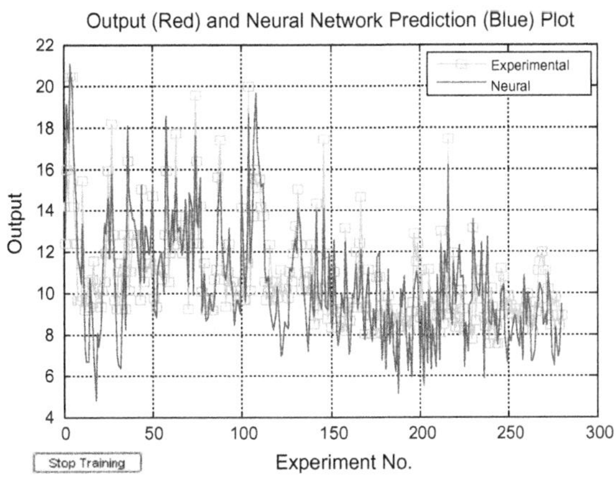

Fig. 8.7 Experimentation and ANN model value of HE

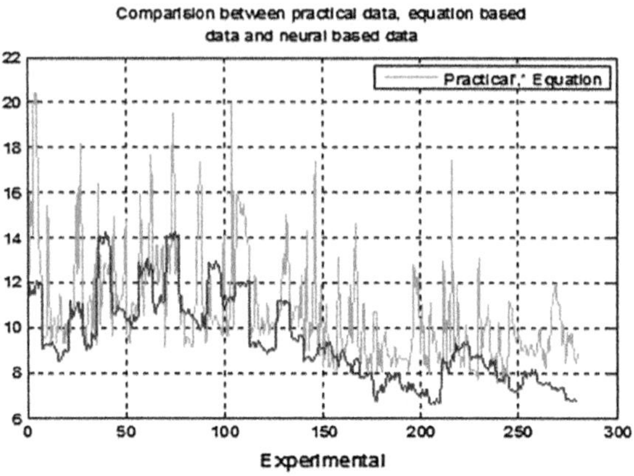

Fig. 8.8 Experimentation and mathematical model value of HE

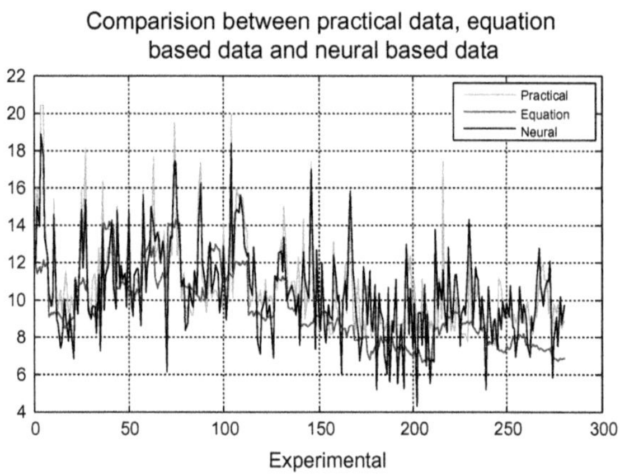

Fig. 8.9 Experimentation, ANN and mathematical model of HE

mathematical model is lower compared to earlier two methods. Vibration transmission affects the pulse rate, and hence, there is difference in the values of human energy consumption by mathematical model and experimental value.

8.6.1 Performance Analysis of Drilling Operation Model

Human energy consumption obtained by mathematical equation, ANN model and experimental values is indicated in Table 8.4. The mean absolute error is 1.1898 because all these values match very well with each other. The less percentage error gives an accurate relationship between ANN simulation and experimental data.

Table 8.4 Experimental, ANN and mathematical model values of HE

Dependent π term	π_{D1} Energy
Mean Field	10.7704
Mean ANN	10.4641
Mean Model	9.4520
Mean absolute error performance function	1.1898
Mean squared error performance function	2.3679

8.7 Conclusion

The experimentation data is analysed with respect to human energy consumption by operator during drilling operation with respect to posture angle of operator. Through sensitivity analysis, the most dominant parameter in the consumption of human energy is identified. Human energy consumption obtained by mathematical equation, field values and ANN prediction model is nearly same. Human energy values obtained by these methods followed the same trend, i.e., they are overlapping to each other. Human energy value calculated by mathematical equation is less compared to remaining two methods. Regression analysis also gave the mapping of actual and predicted values of human energy which shows R value 0.998. ANN model is validated with negligible error. Mathematical model, ANN prediction and field values for human energy truly represent the degree of interaction of various independent variables. The mean error of estimate of values of dependent variable is found to be very low. This gives authenticity to mathematical model and ANN model. In sensitivity analysis, the dominant factor in mathematical equation consists of various factors like age, weight, posture angle, hardness of material and operational specifications is identified. Among these factors, age of the operator is observed more dominant for the human energy consumption.

References

Dong, J.H., Dong, R.G.: A method for analyzing absorbed power distribution in the hand and arm substructures when operating vibrating tools. J. Sound Vib. **311**, 1286–1304 (2008)

Dong, R.G., Welcome, D.E.: Measurement of biodynamic response of human hand arm system. J. Sound Vib. **294**, 807–827 (2006)

Gerhardsson, L., Balogh, I.: Vascular and nerve damage in workers exposed to vibrating tools. The importance of objective measurements of exposure time. Appl. Ergon. **36**, 55–60 (2005)

McDowell, T.W., Wiker, S.F., Dong, R.G.: Evaluation of psychometric estimates of vibratory hand-tool grip and push forces. Int. J. Ind. Ergon. **36**, 119–128 (2006)

Neely, G., Burstorm, L.: Gender difference in subjective responses to hand arm vibration. Int. J. Ind. Ergon. **36**, 135–140 (2006)

Chapter 9
Modelling the Perception Towards In-Vehicle Distracted Driving Among Four-Wheeler Drivers in Kerala

R. Srinath, R. Rajesh, R. Sasikumar and B. Subin

Abstract Every year, a huge number of road traffic crashes occur in Kerala with enormous fatality and injury. The major reported reason for road traffic crashes is driver's error. A common observed reason for this error, though less reported, is distraction caused to the driver. With the increasing accessibility of infotainment devices inside the vehicle, drivers are likely to be distracted from driving. The objective of this paper is to model the relationship between perception of four-wheeler drivers in Kerala towards distracting sources/activities and associated safety risk concern. The distracting sources and activities among four-wheeler drivers were identified through a literature review. A self-reported survey was conducted in Kerala (n = 1203) using a four-part questionnaire administered by random sampling approach in offline and online modes. It consisted of questions relating to exposure to driving distractions, perception towards distraction from different sources and activities, perceived risk of incidents due to distraction and suggestions for deriving engineering and administrative solutions for the improvement of safety. A confirmatory factor analysis followed by structural equation modelling was done to understand the direct and indirect effects on safety perception of drivers towards causes and consequences of in-vehicle distracted driving. The study highlights the need to concentrate on driver–vehicle interactions that cause in-vehicle distractions and safety incidents. Further, the study calls for focusing on scientific training methods and technological interventions that would address the issue of in-vehicle distraction.

Keywords Distracted driving · Mobile phone · Safety perception Survey · Structural equation modelling

---

R. Srinath (✉) · R. Rajesh · R. Sasikumar · B. Subin
National Transportation Planning and Research Centre, Thiruvananthapuram, India
e-mail: krishna.srinath@gmail.com

J. Maiti and P.K. Ray (eds.), *Industrial Safety Management*,
Managing the Asian Century, https://doi.org/10.1007/978-981-10-6328-2_9

9.1 Introduction

Every year, a huge number of road traffic accidents are reported in Kerala. A major portion of the accidents occurs due to the fault of drivers. An important reason for the drivers to commit faults is the momentary loss of attention and control of the vehicle. Driving under conditions in which a driver loses his attention momentarily resulting in crashes, near misses, lane deviations, changes in driving speed, etc., is known as distracted driving. Driving distractions can occur due to environmental sources inside the vehicle such as improper lighting, position of instruments and equipments, uncomfortable seating, etc., and behavioural aspects related to execution of some tasks in parallel to driving such as interaction with passengers, eating or drinking while driving, manipulating controls of electronic equipments and adjusting the environment inside the vehicle, use of mobile phone, etc. As a result, the driver may not be able to address emergency situations.

Driver distraction is a kind of driver inattention that occurs when drivers divert their attention from the driving task and allot the same for some other non-driving activity (NHTSA 2015). There is a huge volume of literature focused on the use of mobile phones while driving (Atchley et al. 2011; Caird et al. 2008, 2014; Nelson et al. 2009; Rajesh et al. 2016; Cuenen et al. 2015). The attention capacity has a moderating effect on the stable control of the drive, and older drivers choose compensating methods at distracting situations (Cuenen et al. 2015). Knapper et al. (2015) has investigated the distracting effect of smart phones and navigation systems on driving in urban and other motorways with the help of a simulator system. Driver distractions have grown up with the development in sophisticated technology, the most critical distraction being the cell phone (Brodsky and Slor 2013). Text messaging poses greater risk as compared to talking because of its requirement for physical commitment of the driver. (Brodsky and Slor 2013) and (Ünal et al. 2012) have analysed the influence of background music as a risk factor for distraction among drivers. Carter et al. 2014, based on study that measures socially-oriented, task-oriented and environment-oriented behaviours while driving, states that perception towards safety risk is the key predictor of distracted driving behaviour. According to Golias et al. (2002), the environmental parameters inside the vehicle such as temperature, seating, rear and side mirrors, etc., must be identifiable and adaptable for the driver, otherwise adjusting them will be distracting and it compromises safety.

From the literature review, the internal factors that can potentially distract the driver from safe operation of the vehicle have been identified. They are conversing with a passenger, use of mobile phone, eating/drinking while driving, use of navigation aids, background music and adjusting radio/CD player. No study relating to distracted driving is seen to be undertaken in India, and no published articles have been found in this regard. It is also identified that specific data pertaining to distracted driving incidents in accident reporting and recording systems in the state is absent.

9.2 Problem Statement

The use of infotainment systems and devices is increasing significantly among car drivers. With the escalated involvement in secondary tasks, the likelihood of distraction while driving is also assumed to go high. To address the problem of distractions, it is necessary to understand the relationship between distractions realized or perceived by drivers and the level of risk perceived while undertaking secondary tasks and being distracted.

9.3 Objective

The objective of this paper is to model the relationship between perception of four-wheeler drivers in Kerala towards distracting sources/activities and associated safety risk concern. Based on the study, a few remedial measures to the problem are suggested.

9.4 Methodology

A self-reported qualitative survey was conducted in Kerala (n = 1203) using a four-part questionnaire derived from a literature review and administered by random sampling approach in offline and online modes. Earlier, a pilot survey was conducted (n = 110) in order to check the reliability of the questionnaire. Part A consists of questions relating to exposure to driving distractions. Part B consists of questions on perception towards distraction from different sources and activities. Part C comprises of questions pertaining to perceived risk of incidents due to distraction. Part D includes questions for deriving engineering and administrative solutions for the improvement of safety. Responses for part A questions are on a 5-point Likert-type scale with one being 'Never' and five being 'Regularly'. Part B and C comprises of questions with responses on 5-point Likert scale with one being 'Strongly Disagree' and five being 'Strongly Agree' The hypothetical model comprises of internal environment, driver and drive characteristics, use of mobile phone, visual distractions and non-visual distractions. The most vital component that directly influences safety is the visual sense of the driver. The effect of sources and activities on visual and non-visual distraction is also assessed. The hypothetical model is shown in Fig. 9.1. A confirmatory model (measurement model) is developed and then shaped into a structural equation model (SEM) by connecting the perception towards distracting factors and safety risk. IBM SPSS is used for statistical analyses and IBM SPSS AMOS is used for the structural equation modelling. After verifying for systematic errors, a total of 1051 responses have been chosen for modelling.

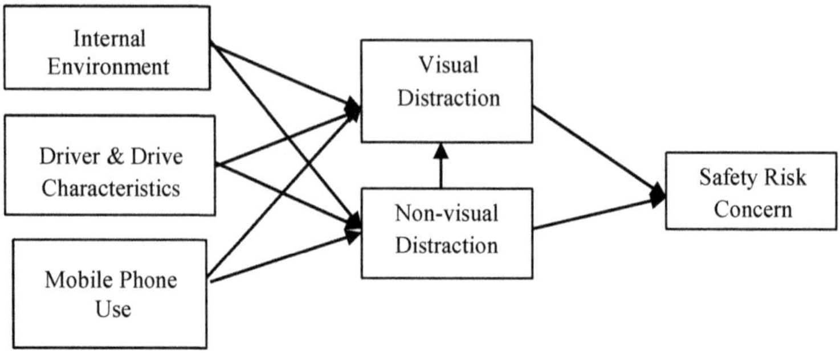

Fig. 9.1 Hypothetical model for distracted driving

9.5 Results and Discussion

The demographic details relating to the respondents are given in Table 9.1. The mean (std dev) of age of the respondents was 33.14 (SD = 10.66). Details of the extent and nature of in-vehicle distractions can be found in (Srinath et al. 2015).

9.5.1 Structural Equation Model

The measurement model of in-vehicle distracted driving is given in Fig. 9.2, whose model fit parameters are $\chi 2 = 394.443$(df = 125), p = 0.000, root mean residual (RMR) = 0.043, goodness-of-fit index (GFI) = 0.959, adjusted goodness-of-fit index (AGFI) = 0.943, comparative fit index (CFI) = 0.931, root mean square error of approximation (RMSEA) = 0.045. The measurement model is used to analyse the interrelationship between causative factors. The measurement model is connected to the effect factor, which in this case is the safety risk concern due to different in-vehicle distracting sources and activities. This forms the initial structural model shown in Fig. 9.3. This model is tested for fitness, and modification is made in single iterations by adding error covariances and removing insignificant relationships between factors. Covariances are added between error variables which show the highest modification index. The modification history of the initial structural model is given in Table 9.2. The regression coefficients for the paths in the distraction model (Fig. 9.4) obtained through the maximum likelihood method are shown in Table 9.3. The critical ratios for estimates of covariance between the constructs and variance of indicators are observed to be greater than 1.96 and $p < 0.05$. The final structural equation model developed for in-vehicle distracted driving in Kerala is shown in Fig. 9.4.

In this study, it has been found that in the in-vehicle environment factor the travel duration has more influence (0.62) than other sources such as closed environment, audibility inhibition due to background music and driving speed. Use of

Table 9.1 Descriptive statistics of the sample

Age group				Gender		Experience			
18–30	31–40	41–50	Above 50	Male	Female	< 5 years	5–10 years	10–15 years	>15 years
50.59%	23.44%	18.33%	7.54%	84.8%	15.2%	41.97%	32.83%	13.96%	10.05%
Occupation		Driving speed				Penalized		Met with accident	
Professional driver	Private driver	Below 40 kph	40–60 kph	60–80 kph	Above 80 kph	Yes	No	Yes	No
17.62%	82.38%	9.64%	65.5%	21.6%	1.5%	15.62%	84.04%	16.04%	83.96%

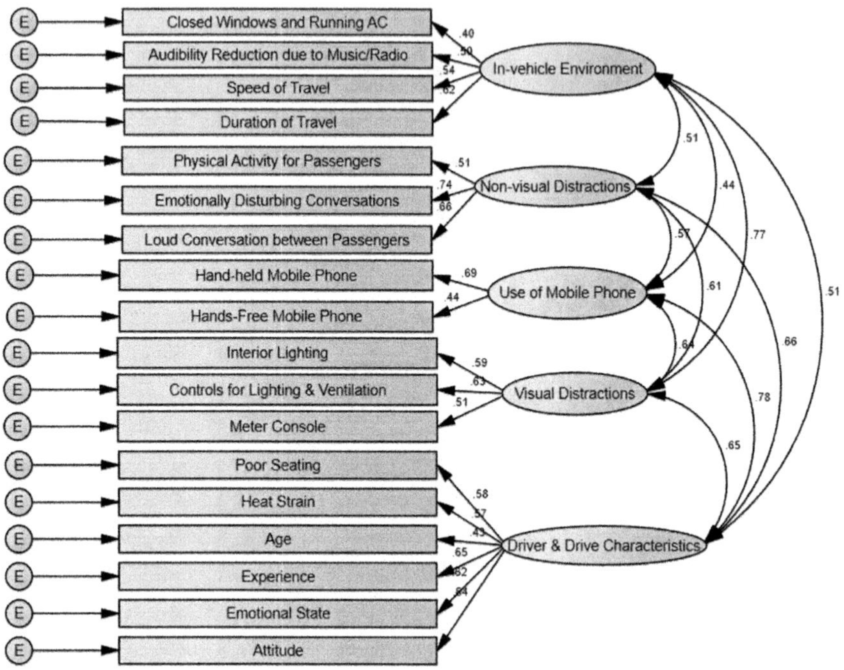

Fig. 9.2 Measurement model for the in-vehicle distracted driving in Kerala

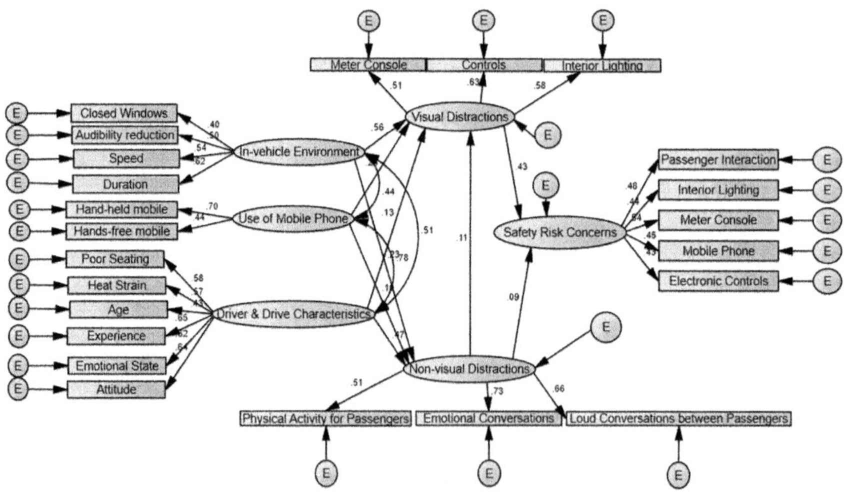

Fig. 9.3 Initial structural model for in-vehicle distracted driving in Kerala

Table 9.2 Modification history of structural model of in-vehicle distracted driving in Kerala

Model	Modification	χ2 (df), p	RMR	GFI	AGFI	CFI	RMSEA
Default	None	830.59(219), 0.000	0.057	0.934	0.916	0.871	0.052
Change 1	Error covariance, Interior Lighting-Mobile phone, MI = 51.991	751.24(218), 0.000	0.053	0.940	0.924	0.888	0.049
Change 2	Error covariance, Metre console-Electronic controls. MI = 31.159	715.42(217), 0.000	0.051	0.943	0.927	0.895	0.047
Change 3	Error covariance, Mobile phone-Electronic controls. MI = 34.172	659.85(216), 0.000	0.048	0.946	0.931	0.907	0.044
Change 4	Remove 'Use of Mobile Phone', 'Non-visual distraction' interaction, insignificant p value	660.66(217), 0.000	0.048	0.946	0.931	0.907	0.044

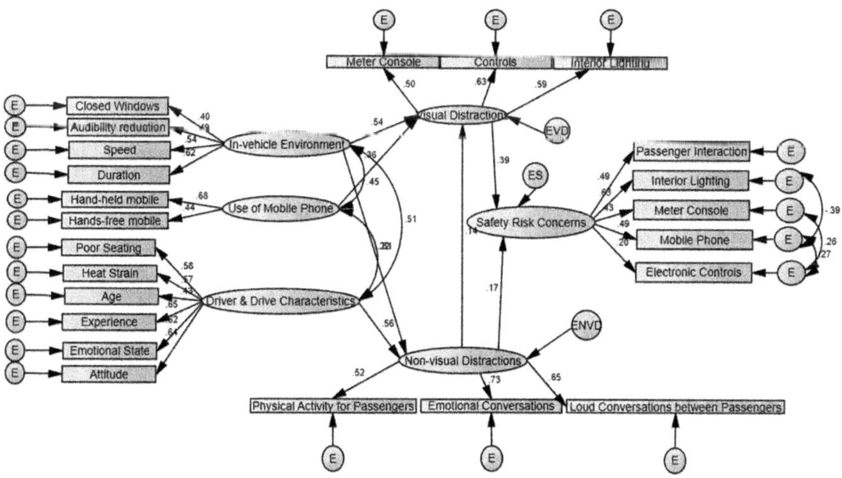

Fig. 9.4 Final structural model for in-vehicle distracted driving in Kerala

handheld mobile phone is perceived as more harmful (0.68) than hands-free phones (0.44). Among the driver characteristics, experience (0.65), emotional state (0.62) and attitude (0.64) have been rated to be of higher influence as compared to age

Table 9.3 Regression weight estimates for the structural paths of distracted driving model

			Standardized regression coefficient	CR	p
Closed Windows	←	In-vehicle Environment	0.399	9.732	***
Audibility reduction	←	In-vehicle Environment	0.491	11.388	***
Speed	←	In-vehicle Environment	0.543	12.158	***
Duration	←	In-vehicle Environment	0.622		
Handheld mobile	←	Use of mobile phone	0.677	10.619	***
Hands-free mobile	←	Use of mobile phone	0.441		
Poor seating	←	Driver and drive characteristics	0.579	15.302	***
Heat strain	←	Driver and drive characteristics	0.570	15.117	***
Age	←	Driver and drive characteristics	0.427	11.803	***
Experience	←	Driver and drive characteristics	0.651	16.744	***
Emotional state	←	Driver and drive characteristics	0.620	16.149	***
Attitude	←	Driver and drive characteristics	0.638		
Metre Console	←	Visual distraction	0.503		
Controls	←	Visual distraction	0.630	12.593	***
Interior lighting	←	Visual distraction	0.593	12.242	***
Physical activity for passengers	←	Non-visual distraction	0.516	13.080	***
Emotional conversations	←	Non-visual distraction	0.733	15.999	***
Loud conversation between passengers	←	Non-visual distraction	0.654		
Passenger interaction	←	Safety risk concerns	0.487		
Interior lighting	←	Safety risk concerns	0.626	9.155	***
Metre console	←	Safety risk concerns	0.433	9.483	***
Mobile phone	←	Safety risk concerns	0.486	8.074	***
Electronic controls	←	Safety risk concerns	0.197	4.635	***

(0.43) as distracting factors. Control of lighting and ventilation (0.63) causes more visual distraction compared to internal lighting (0.59) and position of metre console (0.5). Emotionally disturbing conversations (0.73) have more effect on non-visual distractions than attending to the needs of passengers (0.52) and conversation between passengers (0.65). The effect of in-vehicle environment on visual distraction is found to be more (0.54) as compared to that due to the use of mobile phones (0.45). 'Driver and Drive Characteristics' are related to non-visual distraction (0.56), and its relation with visual distraction has been found to be insignificant and hence is removed. The interaction between visual and non-visual distractions has been found to be feeble; however, effect of multiple secondary tasks can be studied only from experimental studies. The effect of visual distractions on safety risk concerns (0.39) is found to be more than the effect due to non-visual distractions (0.17). However, this implies that there is a reduced level of perception towards overall risk associated with in-vehicle distracted driving among the subjects. The variables in the safety risk concerns factor show similar regression weights except in the case of risk due to entertainment systems (0.2) such as radio, stereo, etc. The distraction due to passengers (0.49), internal light source (0.53), frequent glancing into metre console (0.43) and use of mobile phone (0.49) have moderate effects on safety concerns due to in-vehicle distractions. The significance level (p) of all relationships has been found to be substantially lower than 0.05, and hence, the model can be said to be closely fit.

Horberry et al. (2006) and Simons-Morton et al. (2015) have identified significant differences in young and old drivers in undertaking secondary distracting activities. Shinar et al. (2005) has stated that as experience builds up, drivers get more used to the routine task of driving and they get involved in additional activities. Briggs et al. (2011) has deduced that the more emotionally drivers are involved in a conversation, the greater potential for distraction exists. All these studies point out that despite perceiving the risk of crashes and injuries, drivers choose to involve in secondary tasks, mostly using mobile phones.

9.5.2 Recommendations

Engineering and administrative interventions are required to mitigate the problem of in-vehicle distracted driving. The ideal location and type of interior lighting in four-wheeled vehicles have to be determined by proper experimentation and design. Out of the obtained responses, placing the metre console above the dashboard, behind the steering wheel is the most popular suggestion. Drivers can be relieved of glancing into the metre console by providing suitable regulation systems. Sensor-based traffic density measurement can also be developed to determine safe speed limits to enable comfortable and steady drive without incidents. Regulation of internal environment can be automated. Anthropometric designing and automatically adjusting seats must be provided for more comfort, to prevent unnecessary movement and chance of fatigue. Another intervention that can be experimented in four wheelers is the use of paddle gear instead of the gear lever fitted on the floor of the vehicle. For control of infotainment equipments while driving, voice-enabled recognition interfaces and voice level regulating feedback systems have to be developed. The electronic equipments have to be placed at locations where the driver needs to invest the least attention and physical commitment.

There are many types of driver assistance systems that are available in today's automobiles. Tsugawa et al. (2006) defines a driver assistance system (DAS) as a mechanism or system that covers part of the sequence of tasks (recognition, decision making and operation) that a driver performs while driving a motor vehicle. Many driving assistance systems have been proposed and developed (Alghamdi et al. 2012; Gaikwad and Lokhande 2015). Rodriguez-Ascariz et al. (2011) has presented a primary level low-cost, non-interfering, small-sized electronic system for detecting the use of mobile phones by drivers while driving. Cost-effective systems have to be developed in order to make the drivers more aware of the potential threat and prevent traffic incidents.

More technology oriented systems have to be utilized for licensing and monitoring purposes. Awareness of problems due to distracted driving can be imparted using simulator-based driving assessment in which many constraints can be included. Progressive licensing system and continuous monitoring will help to develop skilled and quality drivers from time to time. Traffic surveillance system in the state has to be upgraded to address externally detectable driver faults, and systems for making the driver aware of the faults have to be implemented so that repetition of such act can be prevented.

9.6 Conclusions

Road safety is a growing epidemiological concern in Kerala. Intense restrictions in infrastructure development and drastic growth in vehicle technology create adverse effect on safety and health of individuals as well as on the economy of the state. Reports show that the most detrimental reason for road traffic haphazard is the fault of drivers. An important reason for this fault is the distraction caused to the driver due to some voluntary or involuntary factor. A study on driver's activities inside the vehicle will help to identify potentially dangerous situations and the distracting reasons for such conditions. In order to model the relationship between perception on distracted driving sources and safety risk concerns among four-wheeler drivers in Kerala, a questionnaire survey has been conducted. 1203 responses have been collected, analysed and modelled using structural equation model.

The variables included in the study are classified under in-vehicle environment, driver and drive characteristics, use of mobile phone, visual distractions and non-visual distractions. A structural equation model has been developed with a sample size of 1051 responses. The final structural equation model has model fitness parameters, $\chi 2 = 660.66$ (df = 217), $p = 0.000$, RMR = 0.048, GFI = 0.946, AGFI = 0.931, CFI = 0.907 and RMSEA = 0.044. Key finding of the study is that the level of perception of risk towards distracting sources and activities among drivers ranges from low to moderate. Good and safe-minded drivers should be able to understand the significance of staying focussed and the possible consequences of distraction. Unfortunately, such a safety-mindedness is not prevalent in majority of drivers in the state. Based on the study, a few interventions in engineering and administrative domains have been recommended.

One limitation to this study is the underrepresentation of female and old age drivers. Further scope in this study lies in the experimental evaluation of the model. It can help in deriving objective results and conclusions.

References

Atchley, P., Atwood, S., Boulton, A.: The choice to text and drive in youngerdrivers: Behavior may shape attitude, Accid. Anal. Prev. **43**, 134–142 (2011)

Alghamdi, W., Shakshuki, E., Sheltami T.R.: Context-aware driver assistance system. Procedia Comput. Sci. **10**, 785–794 (2012)

Briggs, G.F., Hole, G.J., Land, M.F.: Emotionally involving telephone conversations lead to driver error and visual tunneling. Transp. Res. Part F: Traffic Psychol. Behav. **14**(4), 313–323 (2011)

Brodsky, W., Slor Z.: Background music as a risk factor for distraction among young-novice drivers. Accid. Anal. Prev. **59**, 382–393 (2013)

Caird, J.K., Willness, C.R., Steel, P., Scialfa, C.: A meta-analysis of the effects of cell phones on driver performance. Accid. Anal. Prev. **40**, 1282–1293 (2008)

Caird, J.K., Johnston, K.A., Willness, C.R., Asbridge, M., Steel, P.: A meta-analysis of the effects of texting on driving. Accid. Anal. Prev. **71**, 311–318 (2014)

Carter, P.M., Bingham, C.R., Zakrajsek, J.S., Shope, J.T., Sayer, T.B.: Social norms and risk perception: predictors of distracted driving behavior among novice adolescent drivers. J. Adolesc. Health **54**, S32–S41 (2014)

Cuenen, A., Jongen, E.M.M., Brijs, T., Brijs, K., Lutin, M., Van Vlierden, K., Wets, G.: Does attention capacity moderate the effect of driver distraction in older drivers? Accid. Anal. Prev. **77**, 12–20 (2015)

Gaikwad, V., Lokhande, S.: Vision based pedestrian detection for advanced driver assistance. Procedia Comput. Sci. **46**, 321–328 (2015)

Golias, J., Yannis, G., Antoniou, C.: Classification of driver assistance systems according to their impact on road safety and traffic efficiency. Trans. Rev. **22**(2), 179–196 (2002)

Horberry, T., Anderson, J., Regan, M.A., Triggs, T.J., Brown, J.: Driver distraction: the effects of concurrent in-vehicle tasks, road environment complexity and age on driving performance. Accid. Anal. Prev. **38**, 185–191 (2006)

Knapper, A.S., Hagenzieker, M. P., Brookhuis, K.A.: Do in-car devices affect experienced users' driving performance? IATSS Res. **39**, 72–78 (2015)

National Highway Traffic Safety Administration(NHTSA): Traffic Safety Facts (2015)

Nelson, E., Atchley, P., Little, T.D.: The effects of perception of risk and importance of answering and initiating a cellular phone call while driving. Accid. Anal. Prev. **41**, 438–444 (2009)

Rajesh, R., Srinath, R., Sasikumar, R., Subin, B.: Modeling Safety Risk Perception due to Mobile Phone Distraction among Four Wheeler Drivers. IATSS Res. **1**, 30–37 (2016)

Rodriguez-Ascariz, J.M., Boquete, L., Cantos, J., Ortega, S.: Automatic system for detecting driver use of mobile phones. Transp. Res. Part C **19**, 673–681 (2011)

Shinar, D., Tractinsky, N., Compton, R.: Effects of practice, age and task demands on interference from a phone task while driving. Accid. Anal. Prev. **37**, 315–326 (2005)

Simons-Morton, B.G., Klauer, S.G., Ouimet, M.C., Guo, F., Albert, P.S., Lee, S.E., Johnathon, P. E., Anuj, K.P., Dingus, T.A.: Naturalistic teenage driving study: findings and lessons learned. J. Saf. Res. **54**, 29–44 (2015)

Srinath, R., Rajesh, R., Sasikumar, R., Subin, B.: In-vehicle Distraction among Four Wheeler Drivers in Kerala. Paper presented at the International Conference on Humanizing Work and Work Environment (HWWE), Bombay (2015)

Tsugawa, S.: Trends and issues in safe driver assistance systems-driver acceptance and assistance for elderly drivers. IATSS Res. **30**(2), 6–18 (2006)

Ünal, A.B., Steg, L., Epstude, K.: The influence of music on mental effort and driving performance. Accid. Anal. Prev. **48**, 271–278 (2012)

Part III
Safety Management

Chapter 10
Assessment of Significance of Safety Life Cycle Management (SLCM) Approach Using IEC61508 Safety Standard Towards Its Implementation in the Foundry Shop: Case Study of a Pump Manufacturing Industry

Kumar Rohit, P.L. Verma and K.K. Ghosh

Abstract The authors have designed and developed a framework to implement the safety life cycle management (SLCM) approach for a safety-instrumented system (SIS) aimed at the overall plant safety and analysed the implementation prospects of safety life cycle (SLC) model while discussing the case of foundry operations in a pump manufacturing industry to minimize the occurrence of different induction hazards (hazards associated with the induction furnace). This study has incorporated the SLC model based on IEC 61508 safety standard comprising the three functional phases—analysis, realization and operation phases, for the safer operation of induction furnace. The SIS's architecture of more diagnostics plus redundancy has to be designed for the increased safety integrity level (SIL) which has been recommended and suggested based on the mathematical interpretations from the study by the computation of safety integrity level (SIL) in conformance with the allocation of safety-instrumented system (SIS) required to overcome the various induction hazards. The increased plant safety, reduced risk and better operational performance, all these can be achieved as the outcome from implementation prospects of safety life cycle management (SLCM) approach. The SLC model being incorporated in this study will gain significant momentum in terms of the overall safety and can benefit the industrial processes and enterprise business as a whole by minimizing the different types of losses in terms of human, material, asset, manufacturing cycle time, etc.

K. Rohit (✉)
IIT Kharagpur, Kharagpur, India
e-mail: rohitkumar.mechy@gmail.com

P.L. Verma
Department of Mechanical Engineering, Samrat Ashok Technological Institute (SATI), Vidisha, India

K.K. Ghosh
Vinod Gupta School of Management (VGSOM), IIT Kharagpur, Kharagpur, India

J. Maiti and P.K. Ray (eds.), *Industrial Safety Management*,
Managing the Asian Century, https://doi.org/10.1007/978-981-10-6328-2_10

Keywords Safety instrumented system (SIS) · Safety life cycle management (SLCM) · IEC 61508 safety standard · Safety integrity levels (SIL)
Safety life cycle (SLC) model

10.1 Introduction

There might be possibility of occurrence of accidental incidents within the factory premises, irrespective of the application of a wide range of safeguarding measures such as use of personal protective equipment (PPE), engineering controls and safe acts/work practices/operating procedures. From the viewpoint of Nair (2011), industrial accidents or hazards remain a major concern for the safety personnel/authorities due to the loss of lives as well as damage to the property and the environment that is inflicted on the society besides upsets in tranquil and heavy economic strain. The efforts are also underway to minimize the damages and ensure all safer working environments around the industrial installations focusing on equipment-based safety. Yadav (2011) addressed that in a country, in any of its economic activities, factors influencing safety issues are mainly socio-economic conditions, infrastructural facilities and technological limitations and most importantly the mindset of people and its human resources. Towards this, the emergence of new era of modern safety management has taken the shape of "safety life cycle management (SLCM)" as its core concept that emphasizes on the strict control of safety-related activities that have been highlighted by Knegtering (2002) and Macdonald (2004). The real challenges lie in managing these safety issues to avoid accidental incidents while keeping the wheels of growth moving by improving performance and profitability. The outline of this study pinpoints the implication perspective of SLC model to address these challenges in order to minimize the occurrences of induction hazards in one of the manufacturing sectors.

10.1.1 Safety Instrumented System (SIS)

As inferred by Ali (2007), the SIS reduces the risk from a hazardous process to a tolerable level by selecting a safety integrity level (SIL).

As stated by Macdonald (2004), this system also traditionally termed as "safety-related systems (SRS)" comes under different names being taken in use such as trip and alarm system, emergency shutdown system, safety interlock system. Lundteigen et al. (2009) discussed the significance of SIS as they have extensive applications in many industrial sectors to mitigate the risk of different losses in terms of human lives, environmental aspects and material assets. These authors explained the functional aspects of a SIS as "A SIS is installed to detect and respond to the onset of hazardous events by the use of electrical, electronic, or programmable electronic systems (E/E/PES) technology".

10.1.2 IEC 61508 Safety Standard

Macdonald (2004) defined IEC 61508 safety standard as " … the first international standard issued/published by the International Electrotechnical Commission (IEC) in 1998 towards the modern safety management, as a successor of draft standard IEC 1508 published in 1995 by IEC in the safety circles", which sets out a complete management procedure and design requirements for overall safety control systems and focuses on design and management of the functional safety systems, technical safety requirements, personnel's competence and documentation. From the literature support of Redmill (1999), it covers a wide range of activities and equipment associated with functional safety designed to cater to rapidly developing technology, encapsulating full life cycle concepts, which can be applied to the safety systems using Electrical/Electronic/Programmable Electronic Systems (E/E/PES), for example relays, instruments, networks.

10.1.3 Safety Integrity Level (SIL)

Ali (2007) conceptualized the safety integrity level (SIL) as "The system's greater importance to safety lowers the rate of unsafe failures" and this gives the measure of the rate of unsafe failures as the safety integrity of the system which in turn defined as, 'the likelihood of a safety-related system satisfactorily performing the required safety functions under all the stated conditions within a stated period of time'.

As defined in the IEC 61508 safety standard (IEC 61508, 1998), SIL is a discrete level (1of 4) for specifying the safety integrity requirements of safety functions. SIL is a target probability of dangerous failure of a defined safety function. Different SILs, according to the high (continuous) demand mode of operation expressed in terms of probability of dangerous or unsafe failure per hour and the low demand mode of operation measured as the probability of failure to perform its safety functions on demand within the specified numeric ranges of probability, have been indicated in Table 10.1.

Table 10.1 SILs for various operation models

Safety integrity level (SIL)	High (continuous) demand mode of operation (probability of dangerous or unsafe failure per hour)	Low demand mode of operation (probability of failure to perform its safety functions on demand)
4	$\geq 10^{-9}$ to 10^{-8}	$\geq 10^{-5}$ to 10^{-4}
3	$\geq 10^{-8}$ to 10^{-7}	$\geq 10^{-4}$ to 10^{-3}
2	$\geq 10^{-7}$ to 10^{-6}	$\geq 10^{-3}$ to 10^{-2}
1	$\geq 10^{-6}$ to 10^{-5}	$\geq 10^{-2}$ to 10^{-1}

Source (Macdonald 2004)

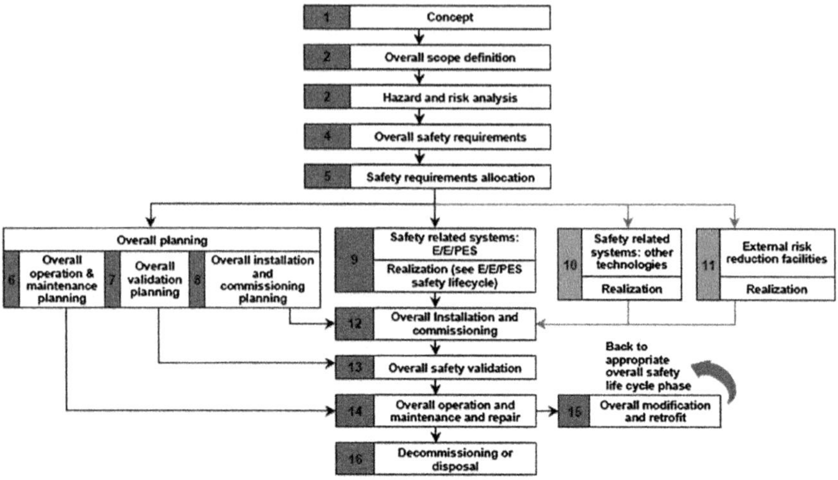

Fig. 10.1 IEC 615O8 overall safety life cycle (SLC) version/model *Source* (Redmill 1999)

10.1.4 Safety Life Cycle (SLC) Model

According to Ali (2007) and Macdonald (2004), it is a structured and systematic model specifying a logical activity flow visualized by a flow chart diagram comprising the phases recommended for the management of safety functions at each stage of the life cycle. As per the IEC 61508 safety standard, SLC model consists of three broad phases—analysis, realization and operation phases. The analysis phase focuses on identifying hazards and hazardous events, chances of occurrence of these hazardous events, potential consequences and the availability of a layer of protection as well as the need for any SISs and allocated SIL. The realization phase focuses on design and fabrication of SISs, and the operation phases comprise start-up, operation, maintenance, modification and eventual decommissioning of the SISs. As stated by Novak and Treytl (2008), these phases encompass the entire life cycle process of safety system from the concept through decommissioning. The IEC 61508 version of SLC model is represented schematically in Fig. 10.1.

10.2 Motivation

Among the hierarchy of required order of actions towards the control of safety-related activities—elimination over hazard, substitution over hazard, engineering control over hazard (equipment-based safety), administrative control over hazard (safe operating procedures/acts/work practices) and the use of personnel protective equipment (PPE)—major emphasis is given on the use of PPEs which should be opted out at last when all the first four orders of actions as per the hierarchy are not sufficient.

Also, the SLC model approach in the field of industrial safety has been mainly focused on the application areas of process industries (chemical/petrochemical industries, etc.) where chances of occurrence of hazards and hazardous events are more pronounced but not in the other areas of engineering applications such as engineering industries (e.g. melting section/furnaces in foundry shop), energy sectors (boiler operation in power plants) where also accidental occurrences or their possibility is more pronounced.

10.3 Case Study and Problem Formulation

This study figures out the safety issues of a pump manufacturing industry located nearby Indore, India, to overcome the induction hazards (hazards associated with the induction furnace) in its foundry division, which might take place during the foundry operations of variable induction power (VIP) series double-track furnace. Such type of furnaces in foundry operations is more prone to the potential induction hazards. Accident investigation analysis reports reveal that hazardous incidents associated with induction furnace might occur due to the induction hazards such as water/molten metal explosion, furnace eruptions (metal splashing), bridging condition (catastrophic explosion), trapping situation of electrocution/electric shocks, lining failure and heat stress. These induction hazards might result in major/minor injuries or may even cause loss of life along with financial, operational and environmental losses.

This study has been conducted from these viewpoints of SLCM approach based on the IEC 61508 safety standard of SLC model:

(1) Identification and assessment of risk prospects of these induction hazards along with their possible causes with the application of hazard and operability (HAZOP) analysis.
(2) Allocation of overall safety requirements in conformance with these induction hazards that can occur during the foundry operations of VIP series double-track furnace.
(3) Design of SIS against these induction hazards based on the allocation of overall safety requirements, i.e. safety integrity level (SIL).
(4) Implementation prospects of SLCM approach to reduce the possibility of accidental incidents to be resulted from these induction hazards.

10.4 Methodology of the Study and Mathematical Computations

Based on this study, the possibility of day-to-day induction hazards during the foundry operations being encountered that may lead to emergency situations can be better understood and dealt with. The SLC model as per the IEC 61508 safety standard can be implemented into three broad phases as illustrated in Fig. 10.1—analysis phases (1–5), realization phases (6–11) and operation phases (12–16)—to reduce the possibility of occurrence of induction hazards, taking into account the case of the pump manufacturing unit.

10.4.1 Analysis Phases

The induction hazards along with their possible sources of occurrence are identified and defined through using one of the methodologies for hazard studies, i.e. HAZOP analysis. Dunjó et al. (2010) addressed HAZOP as a process hazard analysis (PHA) approach that had been used all over the world for reviewing the system's potential hazards as well as also to examine its operability issues by identifying the impact of any deviations from their design conditions.

The HAZOP analysis is a structured, systematic, qualitative and an inductive brainstorming risk assessment tool/technique for system examination and risk management, investigating how the plant conditions might deviate from the design intent as inferred by Macdonald (2004) and Tripathy (2011). From the viewpoint of Ashish et al. (2016), the HAZOP analysis explores and examines the safety issues that symbolize the potential risks to human/equipment or efficient running of operations, which generates the deviations from safe condition using keywords. This study has laid emphasis to incorporate the HAZOP analysis as such, and according to Dunjó et al. (2010), this technique has been intended to improve the plant safety for the systems that extensively operate at very high temperatures and high pressures subjected to the increased complexity of the processes that possess the similar operability conditions as such that of the case of VIP series double-track induction furnace being stated. Using the HAZOP analysis, the specific potential operational hazards (induction hazards) in the system (VIP series double-track furnace) have been identified and quantified. In the HAZOP analysis applied for the induction hazards, some specific guidewords (no/not, more, less, as well as reverse, part of, before/after, etc.) are combined with certain parameters or elements (temperature, moisture, operate, composition, control, sequence, material size, flow, process, etc.) to generate a possible matrix of deviations, shaping into potential induction hazards as the consequences. The HAZOP analysis to identify and assess the risk prospects of induction hazards and to determine their possible causes of occurrence has been stated in Table 10.2.

Table 10.2 HAZOP analysis for VIP series double-track furnace.

Guideword	Element	Deviation	Possible causes	Consequences
No/not (none)	Sequence/phase	No proper pre-heating done	Lack of operator's skill/ technical fault in pre-heater's operation	Violent water/molten metal explosion
More (more of/higher)	Moisture	High moisture content	Introduction of wet/damp metal into melt charge with no proper pre-heating	Violent water/molten metal explosion
Other than	Transfer	Hot but empty furnace	Charging of sealed drums/containers into a hot but empty furnace	Violent water/molten metal explosion
Before/after	Sequence	Pre-heaters/dryers not properly used prior to charging	Carelessness/lack of operator's skill/technical fault in pre-heater's operation	Violent water/molten metal explosion
More (more of/higher)	Particle/material Size	Dropping large pieces of charge material into melt	Metal scraps of larger sizes received by scrap dealers	Furnace eruptions/metal splashing
Other than	Material composition	Cold charges and additives/sealed scraps/scrap rolls	Proper pre-heating not done/Requisite metal scraps not received	Furnace eruptions/metal splashing
Part of	Control	Part of worker's body caught losing control	Improper attention/failure to stand behind safety lines	Trapping situation
As well as (more than)	Operate	Moving equipment/unsafe acts during operations	Improper attention/lack of proper safety training	Trapping situation
More (more of/higher)	Temperature	Very high temperature of the molten bath	Improper attention during charging of the furnace causing superheating	Bridging/catastrophic explosion
Reverse	Separation	Top portion charge not in contact with bottom portion melt	Wall/bridge formation by presence of slag or impurities	Bridging/catastrophic explosion
Less (less of/lower)	Charge	Less/incompletely discharged capacitors	Contact with the incompletely discharged capacitors	Electrocution/electric shocks

(continued)

Table 10.2 (continued)

Guideword	Element	Deviation	Possible causes	Consequences
As well as (more than)	Operate	Unsafe acts/practices during operations	Contact with electrical conductors/overriding safety interlock switches	Electrocution/electric shocks
No/not (none)	Lining process	No lining	Forgotten the lining process/lining damage during new lining/liner not competent	Lining failure (causing metal penetration in coil to result in explosions)
Less (less of/lower)	Lining thickness	Improper or inappropriate lining thickness	Unskilled work and poor supervision during the line preparation	Lining failure (causing metal penetration in coil to result in explosions)
Reverse	Liner material	Installation of the wrong refractory material for a specific application	Inappropriate lining material/not aware about the lining material/poor supervision	Lining failure (causing metal penetration in coil to result in explosions)
No/not (none)	Addition	No flow of water	Water tank being empty/pump winding burnt/ water tube being chocked/air block in cooling line	Heat stress
Less (less of/lower)	Flow	Less flow of water	Scale formation in water tubes/water leakage from pump or tubes/low water level in tank	Heat stress
Reverse	Process (reverse flow)	Hot water	Failure of water cooling system/pump malfunctioning	Heat stress

After conducting HAZOP analysis to analyse and assess the possibility of said consequences, the need for SIS is to be determined by checking available layers of protection against the induction hazards by SIL determination using quantitative method of layer of protection analysis (LOPA) for the estimation of overall safety requirements using these mathematical computations, stated in Macdonald (2004).

10.4.1.1 Mathematical Computations (To Determine the SILs Using LOPA Method)

Risk frequency (Fnp) = Occurrence possibility/flame-out frequency × Probability of chances of a hazardous event/explosion (e.g. x in y per event) [per year]

Target-protected risk frequency (Fp) = 1/No. of years up to which chances are greater [per year]

Target risk reduction factor (RRF) = Fnp/Fp

Average probability of failure to perform its designed function (PFDavg) = 1/RRF (Source: (Macdonald 2004))

Note: SIL is determined for low demand mode of furnace operations based on the computation of average probability of failure to perform its designed safety function (PFDavg) from the tabular representation of comparative SILs, as per Table 10.1.

The allocation of safety requirements against the various induction hazards has been done to the designated safety-related systems (SRS) of allocated SIL based on the estimation of SIL by LOPA method that has been mathematically computed and specified in Table 10.3.

10.4.2 Realization Phases

Ghosh and Miller (2009) stated the significance of realization phases as to design, develop and engineer the safety systems (SRS/SIS) based on the allocation of safety requirements. The design functions and fabrication of SRS/SIS for the VIP series double-track furnace to overcome the induction hazards, according to Macdonald (2004), have to be performed in accordance with the SRS/SIS's architecture of allocated SIL as per the SIL allocation of safety requirements which has been stated in Table 10.2.

10.4.3 Operation Phases

The operation phases have to be designed and engineered in conformance with the severity of induction hazards focusing on the operational aspects of VIP series double-track furnace. The overall installation and commissioning is to be planned

Table 10.3 Computation of SIL for allocation of SRS/SIS required to overcome the induction hazards

Type of induction Hazards	Risk frequency (Fnp) (per year)	Target-protected risk frequency (Fp) (per year)	Target risk reduction factor (RRF)	Avg. probability of failure to perform its designed safety function (PFDavg)	Type of safety system (SRS/SIS) allocated based on SIL classification
Water/molten metal explosion	1 per 3 years = 0.33	1 per 3000 years = 3.33×10^{-4}	1000	1×10^{-3}	SIL 2
Furnace eruptions (Metal splashing)	12 per year = 12	1 per 500 years = 2×10^{-3}	6000	1.67×10^{-4}	SIL 3
Bridging condition	2 per year = 2	1 per 500 years = 2×10^{-3}	1000	1×10^{-3}	SIL 2
Trapping situation	5 per year = 5	1 per 1000 years = 1×10^{-3}	5000	2×10^{-4}	SIL 3
Electrocution/electric shock	1 per year = 1	1 per 2000 years = 5×10^{-4}	2000	5×10^{-4}	SIL 3
Lining failure	2 per year = 2	1 per 500 years = 2×10^{-3}	1000	1×10^{-3}	SIL 2
Heat stress	15 per year = 15	1 per 200 years = 5×10^{-3}	3000	3.33×10^{-4}	SIL 3

and worked out so that SRSs/SISs and external risk reduction facilities are installed and commissioned in a controlled manner to ensure that the overall functional safety has to be achieved for the VIP series double-track furnace against the induction hazards. The SRS/SIS validation for the assessment of overall functional safety achieved has to be performed through functional safety assessment by the means of practical functional testing/impact analysis/ factory acceptance test (FAT)/ pre-start-up acceptance test (PSAT) for SRS/SIS of allocated SIL as suggested by Macdonald (2004) that will minimize the chances of occurrence of induction hazards. The routine operation of SRS/SIS, in terms of operating procedure, has to be standardized along with its maintenance and repair aspects as well as the overall retrofit/modification has to be facilitated with respect to proposed changes in the system, finally arriving at its eventual decommissioning based on the validation of overall safety.

10.5 Results and Interpretations

Based on SIL classification, the overall safety requirement for SRS/SIS against various induction hazards is allocated towards SIL 2 and SIL 3, as graphically represented in Fig. 10.2. As per the allocation of SRS/SIS towards SIL 2 and SIL 3,

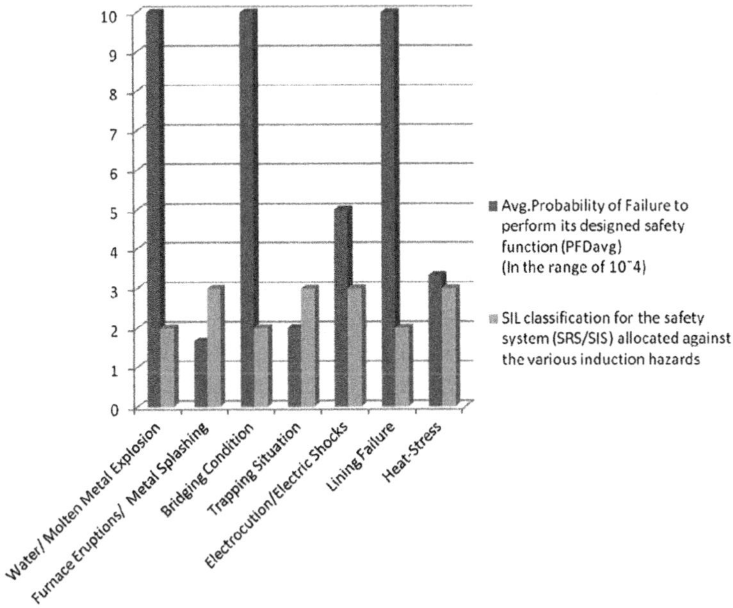

Fig. 10.2 SIL Allocation against various induction hazards

Table 10.4 Architecture of SRS/SIS's design recommended w.r.t. allocated SILs

SIS based on allocated SIL	SRS/SIS's architecture recommended for various allocated SILs
SIL 4	No specific safety-related system (SRS) required
SIL 3	Significant aspects of diverse separation, redundancy and diagnostics
SIL 2	More diagnostics plus redundancy as needed
SIL 1	Acceptable to use single channel architecture

Source (Macdonald 2004)

an increased safety level of SIL 2 has to be opted out focusing on SIS's architecture of more diagnostics plus redundancy to optimize the design of SIS for VIP series double-track furnace, considering overall safety requirements, as interpreted from Table 10.4. The impact will be analysed and updated by any previous hazard and risk assessment. The results of analysis will be used to reactivate relevant requirements including reverification and retesting for VIP series double-track furnace to have a protective layer against the occurrence of induction hazards.

10.6 Conclusions

With the SIS's architecture of more diagnostics plus redundancy designed for the increased SIL, as interpreted from the study, increased plant safety and reduced risk and better operational performance can be achieved towards the safer plant operations. The SIS/SRS's architecture of more diagnostics plus redundancy that met the required increased safety level of SIL 2 will maximize the system availability for VIP series double-track furnace against the induction hazards during the foundry operations and thereby form the basis for the company's new safety system design standards and practices, in terms of overall safety. Safer plant operations, reduced engineering, operation and maintenance costs, increased process uptime as well as decrease in the number of false and unnecessary alarms and nuisance trips with a reduction in their operating costs are all achievable with the implication of SLC model. SLC model can be successfully implemented, but such an implementation together with the control of SLC is only achieved if the emphasis is put on the control of involved safety-related processes. The IEC 61508 version of overall safety life cycle (SLC), if all its phases are successfully and effectively implemented, will continue to play an integral role in the design and evaluation of the safety-related industrial processes.

References

Ali, R.: How to implement a safety life-cycle, Valve Magazine, Valve Manufacturers Association, pp. 1–6 (2007)

Ashish, S., Gerald, L.A., Prakash, R.S.: Comparison of Traditional HAZOP and WHIPP. Appl. Mech. Mater. **854**, 52–58 (2016)

Dunjó, J., Fthenakis, V., Vílchez, J.A., Arnaldos, J.: Hazard and operability (HAZOP) analysis. A literature review. J. Hazard. Mater. **173**(1–3), 19–32 (2010)

Ghosh, A., Miller, P.: Siemens Offers a Comprehensive Product Portfolio for Safety Lifecycle Management, ARC Advisory Group, Dedham (U.S.A.) (2009)

IEC 61508: Functional Safety of Electrical/Electronic/Programmable Electronic Safety-Related Systems. International Electrotechnical Commission, Geneva (1998)

Knegtering, B.: Safety lifecycle management in the process industries: the development of a qualitative safety-related information analysis technique. Technische Universiteit Eindhoven (2002)

Lundteigen, M.A., Rausand,M., Utne, I.B.: Integrating RAMS engineering and management with the safety life cycle of IEC 61508 **94**(12), 1894–1903 (2009)

Macdonald, D.: Practical Industrial Safety, Risk Assessment and Shutdown Systems for Industry. Newness An imprint of Elsevier, Burlington (UK), pp. 1–170 (2004)

Nair, Dr. M.P.S.: Hazardous material storage installations—steps to address concerns on safety and build public confidence. Ind. Saf. Chronicle – Natl. Saf. Coun. **XLII**(3), 24–28 (2011)

Novak, T., Treytl, A.: Functional safety and system security in automation systems—a life cycle model. In: 13th IEEE Conference on Emerging Technologies and Factory Automation. Proceedings of ETFA 2008, pp. 311–31 (2008)

Redmill, F.: An introduction to the safety standard IEC 61508. J. Syst. Saf. Soc. **35**(1), 1–9 (1999)

Tripathy, Dr. D.P.: Safety risk assessment and management in coal mines. Ind. Saf. Chronicle – Natl. Saf. Coun. **XLII**(3), 33–38 (2011)

Yadav, K.S.: Challenges in achieving safety in power plant. Ind. Saf. Chronicle – Natl. Saf. Coun. **XLI**(4), 49–52 (2011)

Chapter 11
Understanding the Human-computer Interaction Behavior in Electrical and Power Systems

Molla Ramizur Rahman

Abstract Simulation has become essential and important in implementing any industrial applications. Various computational methods are involved in simulating real-life practical applications. Such methods find extensive use in electrical and power systems on which this paper focuses. Therefore, evaluation of software has become essential to examine such feasibility. Although ease of use is a crucial factor in software evaluation, factors like simulation time and efficiency are also important in deciding the effectiveness of the software. Such positive and ease of communication within the system improve the industrial competence through workman's satisfaction, and hence, increase productivity. Behavioral pattern between natural human psychologies with artificial computing system needs discussion to develop an optimally feasible system. Such behavior has an impact in deciding the pricing and investment structure from software vendor and its client, respectively. The paper aims in understanding the effectiveness of human-computer interaction by extensively comparing graphical user interface (GUI)-based software with user programming on parameters like time of execution, user friendliness, and accuracy. The study is made on applications related to electrical, electronics and power systems. Discussions are made on the use of such techniques. Investment and cost analysis are also presented. In addition, advantages and disadvantages of this interaction are discussed.

Keywords Computer techniques · Interaction efficiency · Software preferability · Power systems

11.1 Introduction

Simulation is essential in all industrial applications. It is necessary to understand the working and efficiency of any model, well advance before actually building it. It helps to find out the feasibility of the real system. It is important as it takes less time

M.R. Rahman (✉)
Prin. L. N. Welingkar Institute of Management Development & Research, Mumbai, India
e-mail: ramizurscience@yahoo.com

J. Maiti and P.K. Ray (eds.), *Industrial Safety Management*,
Managing the Asian Century, https://doi.org/10.1007/978-981-10-6328-2_11

than building the actual prototype and even saves money. Industrial safety plays a major factor for simulation. It identifies the probable safety measures and requirements that need to be kept in mind before the actual project begins. Simulation plays a vital role in electrical and electronics systems. Thus, it is required to develop quality software for simulation which not only gives high efficiency but also provides flexible human-computer interaction.

Human-Computer Interaction (HCI) in electrical and power systems essentially means to understand the interface between various electrical and power system applications in the industry with the computers. As electrical and power system is complex, it is important to learn the behavioral pattern of such systems for its interaction with humans. Such behavioral understanding would help us to improve the performance of the overall system. This is required to examine on parameters like time, user-friendliness, and accuracy with respect to HCI.

There are various issues and problems faced in the electrical and power system applications. Due to the complex nature of such applications, stresses among the human interacting with the computers are observed greatly. This depends on the nature of the computer programming in the systems. Further, time to resolve the problems varies among the individuals. This leads to the study of time and accuracy in this paper.

Various techniques are involved for interacting with the computer in order to make an efficient electrical and electronics system. It broadly plays a major role in both the domains of electrical and electronics engineering. In electrical engineering, its use is extensive in power systems simulation, fault diagnosis, understanding and modeling circuit, power flow in various systems, and even handling the entire management operations. Power system mainly covers generation, transmission, and distribution of power. Designing of electrical machines is also an important application for which software is used. In electronics engineering, its application is more vast and essential where it is used right from circuit designing to microvolt electronic operations. The paper by Rahman (2016a) highlighted that uses of software in electrical system is average.

It becomes necessary to develop proper software, where it takes into account the user friendliness for human interaction with computer. Increase in ease of work for the employees' due to such HCI increases employee retention rate in an organization. It even leads to customer satisfaction, as it improves the efficiency of persons operating on such systems due to flexibility leading to quality products and services. In domains like electrical and electronics, HCI is crucial due to the complexity and accuracy of the work associated to this sector. Thus, it is essential to have an effective HCI for such sectors.

Again, to develop such flexible software, it is important to understand the behavioral pattern between human and artificial intelligence. Companies developing such software keep in mind behavioral pattern and try to make it flexible, which definitely increases the demand of such software. HCI also has an impact on the cost of software. The paper highlights such concept of cost and investment. It is necessary to have a note of all the parameters, which might affect such interaction.

This paper also makes an attempt to understand "time" affecting HCI in accuracy and user-friendliness using statistical test.

There are various ways to measure the efficiency of HCI. Various software and computer techniques are involved in simulating electrical and electronics system. They might be programming methods or by using GUI. HCI has wide range of descriptions. It is how individuals use devices which might lead to better performance (Carroll 2009). It is important to understand the behavior of human and computer. Thus, this paper aims to understand the human and computer interaction for GUI and programming methods in electrical and electronics applications. HCI deals with the way and measurements of flexibility of human interaction with computers. The knowledge of various other disciplines like psychology, sociology is essential for understanding human-computer interaction as well (Preece and Rombach 1994).

11.2 Literature Survey

It is important to understand human behavior and his reaction in human-computer interaction. The paper by Newell and Card (1985) deals with the psychology involved in human-computer interaction. Human psychology helps to design effective system by understanding the way human behaves, while dealing with various artificial systems. The author Carroll (1997) puts forward that human-computer interaction is the cross section of computer science and psychology. Various efforts are being made to increase the flexibility for human-computer interaction.

Various methods are developed for better HCI. Such techniques are required to increase the effectiveness of the overall system performance. The paper Pavlovic et al. (1997) discusses hand gestures recognition which can be used as an alternative for HCI. To improve HCI, it is necessary to understand the dynamic interaction (Hollan et al. 2000). Designing such interactive techniques depends upon the nature of the work. Interactive methods for teacher and student with computer vary than that of used by the operators in electrical and power systems.

The paper identifies two channels of interaction, which are those of transmitting explicit and implicit messages (Cowie et al. 2001). HCI plays a key role in learning and educational domains. It is important as the students who are involved in the learning process needs to understand the concepts which are new to them. For this, it becomes essential to develop such a system of learning where electronics–computer systems are involved along with students and teachers in the learning environment. Efficiency of the whole system depends upon the interaction effectiveness of the artificial systems with both students and teachers. The paper by Fischer (2001) deals with high functionality applications and user modeling to make a better learning environment. Further, the work by Rahman (2016b) indicated effectiveness of digital education over traditional learning, which can also be used as an evidence to highlight user-friendliness of HCI in computerized digital education.

Therefore, it is important to quantify and measure users' satisfaction of HCI. The study by Chin et al. (1988) shows an effort to understand such user satisfaction by developing questionnaire, where the questionnaire's reliability was observed to be high.

This paper also measures effectiveness in similar line with parameter as user friendliness. The study of user-friendliness is subjective and varies from person to person. Moreover, it becomes essential to evaluate the human-computer interaction in existing system from time to time. In the paper of Jaimes and Sebe (2007), different multi-modal human-computer interactions are reviewed.

The present study compares graphical user interface (GUI) and code-based programming used in electrical and power systems with respect to parameters—time of interaction, accuracy, and user-friendliness. It is also investigated whether time plays an important factor in deciding accuracy and user-friendliness for both GUI and code-based programming. The paper also discusses on investments and cost regarding GUI and code-based programming.

11.3 Research Methodology

Programs related to electrical and electronic applications are considered. Eighteen programs are selected. Out of which, nine was to be solved using GUI and remaining nine using computer programming language. In each set of nine programs, classification was made into simple, average, and complex with three programs in each of the difficulty levels. Such programs were given to programming experts for both GUI and code-based programming. Time taken to execute the programs is obtained and noted in seconds. Further accuracy is obtained in percentage. User friendliness is evaluated by questionnaire study in percentage. Figure 11.1 depicts the flow of research methodology.

One-way ANOVA test is conducted using SPSS software to analyze whether time taken by the person interacting with the computer plays significant role in determining the accuracy and user-friendliness for both GUI and self-programming in all the three difficulty levels. Time in seconds is scaled down by 5. For accuracy and user-friendliness, scale of 1 is assigned with values for 96–100%, 2 for 91–95%, and 3 for below 91%.

11.4 Results and Discussion

In the experiment, eighteen programs were selected. The nature of the programs had three levels of complexity—simple, average, and complex. Thus, eighteen programs were distributed among eighteen individuals, who were asked to formulate and execute the programs related to the electrical and power systems and time of interaction, accuracy, and user-friendliness were noted.

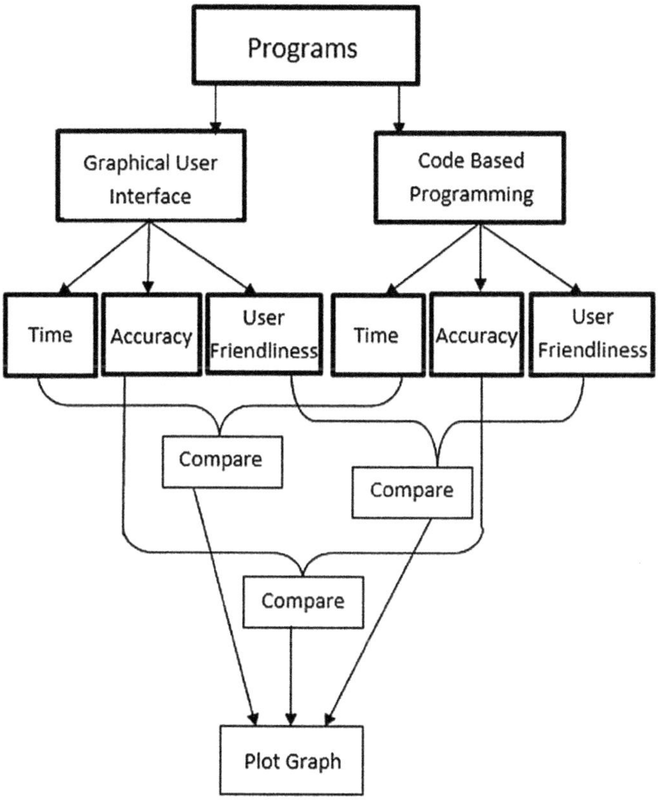

Fig. 11.1 Flowchart of proposed research methodology

Table 11.1 indicates the performance of GUI when compared to programming. The time required for executing programs in GUI is far less when compared to programming in all the three levels of difficulty. Accuracy percentage is high for GUI in all three difficulty levels when evaluated against code-based programming. Similarly, findings are observed for the parameter "user-friendliness".

Figure 11.2 shows the graphical representation of GUI and code-based programming for time. Levels of difficulty are plotted on X-axis, and the time in seconds is plotted on Y-axis. It is evident that GUI-based interaction is better when compared to code-based programming, as the GUI curve lies below code-based programming. Here, less time of interaction is preferred.

Figure 11.3 indicates the graph between levels of difficulty on X-axis and accuracy measured in percentage on Y-axis for parameter accuracy. The curve for GUI lies above the code-based programming indicates that GUI-based programming is better than code-based programming for all three levels of difficulty. Here, higher accuracy is desired.

Table 11.1 Primary study indicating the performance of GUI and code-based programming at different modes of difficulty

Difficulty levels	GUI			Code-based programming		
	Time (s)	Accuracy (%)	User-friendliness (%)	Time (s)	Accuracy (%)	User-friendliness (%)
Simple	3	99	100	9	95	93
Average	6	97	98	14	91	81
Complex	15	95	95	31	85	57

Fig. 11.2 Time study for GUI and self-programming at three levels of difficulty

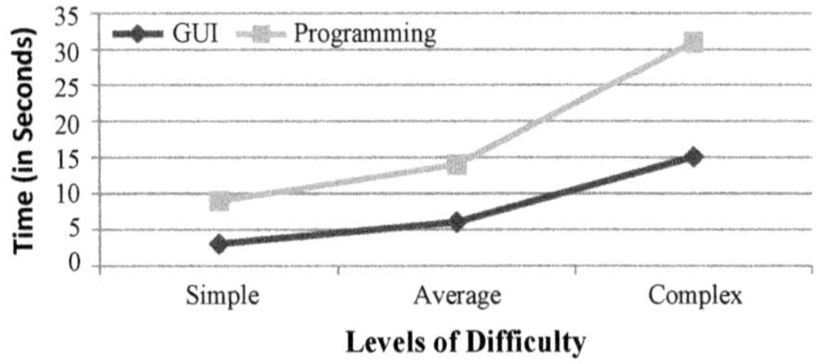

Fig. 11.3 Accuracy of results for GUI and self-programming at three levels of difficulty

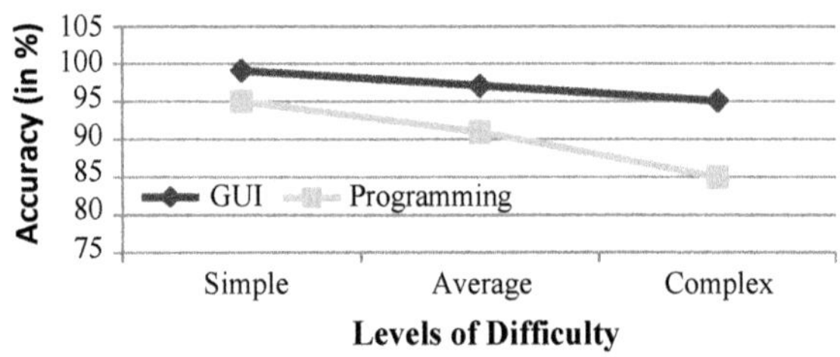

Fig. 11.4 User friendliness for GUI and self-programming at three levels of difficulty

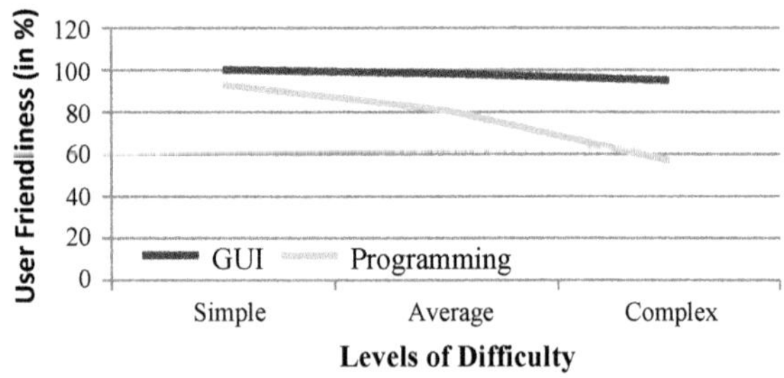

Figure 11.4 indicates the curves for GUI and code-based programming for the parameter user-friendliness, where level of difficulty is plotted on X-axis and user friendliness in percentage is plotted on Y-axis. GUI-based interaction performs better when compared to code-based programming as GUI curve lies above the

Table 11.2 ANOVA test highlighting significance of time with accuracy and user-friendliness

ANOVA						
		Sum of Squares	DOF	Mean Square	F	Sig.
Accuracy	Between Groups	2.333	3	0.778	3.111	0.253
	Within Groups	0.500	2	0.250		
	Total	**2.833**	**5**			
User-friendliness	Between Groups	3.000	3	1.000	2.000	0.350
	Within Groups	1.000	2	0.500		
	Total	**4.000**	**5**			

Note DOF Degrees of freedom

curve for code-based programming for all the three levels of difficulty. Here, higher user friendliness is preferred.

ANOVA test is performed to find whether the factor time has any significance in deciding accuracy and user-friendliness. For this, null hypothesis is formed stating no significance between time and the observed factors, i.e., accuracy and user-friendliness. Hypothesis is formed to find the significance of time with each of the factors accuracy and user-friendliness. Therefore, hypotheses can be shown as:

H0 = There is no significance between time and accuracy.

H0 = There is no significance between time and user-friendliness.

Results of one way ANOVA test are produced in Table 11.2 highlighting the acceptance of null hypothesis, i.e., the time taken by the person on interacting with computer for both GUI and self-programming does not play a factor in determining the accuracy and user-friendliness, as the significance values for accuracy and user-friendliness are 0.253 and 0.350, respectively, which is higher than 0.05.

11.5 Investment and Cost Analysis

Investments for GUI software are higher when compared to self-programming software. But training and expert programmers are required to develop programs in self-programming software, which might have an additional expenditure to the company.

11.6 Conclusion

Human-Computer interaction for electrical and electronics application is studied for both GUI and self-programming. The study is made at three difficulty levels for parameters such as time, accuracy, and user-friendliness. Time for execution of programs at simple, average, and complex levels are 3, 6, and 15 s for GUI, whereas 9, 14, and 31 s for self-programming. Accuracy level of results obtained for GUI is 99, 97, and 95% when compared to 95, 91, and 85% in self-programming at three difficulty levels of simple, average, and complex, respectively. Further user friendliness for GUI was found to be 100, 98, 95% against 93, 81, 57%, respectively, for self-programming at the three above-mentioned levels of difficulty. Further one-way ANOVA test was conducted to understand time as a factor in deciding accuracy and user friendliness. The test result indicated that time has no significant relationship with accuracy and user-friendliness for both GUI and self-programming.

The paper restricts its study to three parameters—total interaction time, accuracy of results, and user-friendliness to compare graphical user interface programming with code-based programming. There are several other parameters which can be studied to understand the interaction behavior. Such parameters can bring out different perspective in understanding human-computer interaction behavior. It is also important to frame out plans to make code-based programming much better with respect to the above-mentioned parameters. Further, it is essential to understand human-computer interaction effectiveness even within various code-based programming languages. Again, the study can be done for understanding the relationship that exists between different parameters of human-computer interaction.

References

Carroll, J.M.: Human-computer interaction: psychology as a science of design. Annu. Rev. Psychol. **48**(1), 61–83 (1997)

Carroll, J.M.: Human–computer interaction. Encyclopedia of Cognitive Science (2009)

Chin, J.P., Diehl, V.A., Norman, K.L.: Development of an instrument measuring user satisfaction of the human-computer interface. In: Proceedings of the SIGCHI conference on Human factors in computing systems, pp. 213–218. ACM (1988)

Cowie, R., Douglas-Cowie, E., Tsapatsoulis, N., Votsis, G., Kollias, S., Fellenz, W., Taylor, J.G.: Emotion recognition in human-computer interaction. IEEE Signal Process. Mag. **18**(1), 32–80 (2001)

Fischer, G.: User modeling in human–computer interaction. User Model. User-Adap. Inter. **11**(1–2), 65–86 (2001)

Hollan, J., Hutchins, E., Kirsh, D.: Distributed cognition: toward a new foundation for human-computer interaction research. ACM Trans. Comput. Hum. Inter. (TOCHI) **7**(2), 174–196 (2000)

Jaimes, A., Sebe, N.: Multimodal human–computer interaction: A survey. Comput. Vis. Image Underst. **108**(1), 116–134 (2007)

Newell, A., Card, S.K.: The prospects for psychological science in human-computer interaction. Hum. Comput. Inter. **1**(3), 209–242 (1985)
Pavlovic, V.I., Sharma, R., Huang, T.S.: Visual interpretation of hand gestures for human-computer interaction: a review. IEEE Trans. Pattern Anal. Mach. Intell. **19**(7), 677–695 (1997)
Preece, J., Rombach, H.D.: A taxonomy for combining software engineering and human-computer interaction measurement approaches: towards a common framework. Int. J. Hum. Comput. Stud. **41**(4), 553–583 (1994)
Rahman, M.R.: Analyzing the performance of a software and IT growth with special reference to India. In: Information Systems Design and Intelligent Applications, pp. 591–598. Springer, India (2016a)
Rahman, M.R.: Comparative study between digital learning and traditional learning: statistical nature and its mathematical valuation. In: Proceedings of the 10th Annual International Conference on Digital Business New Frontiers in Management, pp. 55. Emerald India (2016b)

Author Index

J. Maiti and P.K. Ray (eds.), *Industrial Safety Management*,
Managing the Asian Century, https://doi.org/10.1007/978-981-10-6328-2

Subject Index

J. Maiti and P.K. Ray (eds.), *Industrial Safety Management*,
Managing the Asian Century, https://doi.org/10.1007/978-981-10-6328-2

Printed by Printforce, the Netherlands